The forefront of Korean architecture,
UNSANGDONG

한국 건축의 최전선, 운생동

ARCHI-LAB

UNSANGDONG Architects
Producing

Address.　41, Changgyeong Palace 43Rd., Seongbuk-gu, Seoul, Republic of Korea
Web.　www.usdspace.com

ARCHI-LAB
Publishing

Publisher.　Baeyeon Cho
Address.　18, Yangjaecheon-ro 13-gil, Seocho-gu, Seoul, Republic of Korea
Tel.　+82. 2. 579. 7747
E-mail.　1979anc@naver.com

MasilWIDE
Edited and Designed

Graphic Design.　Den(디이엔), amyyaap
Address.　1F, 45-8, World Cup-ro 8 gil, Mapo-gu, Seoul, Republic of Korea
Tel.　+82. 2. 6010. 1022
Fax.　+82. 2. 6007. 1251
E-mail.　masil@masilwide.com
Web.　www.masilwide.com

Published Date.　19. 10. 2018
ISBN.　979-11-958268-2-7
Price.　34,000won

The forefront of Korean architecture,
UNSANGDONG

한국 건축의 최전선, 운생동

ARCHI-LAB

Contents

006	Any architecture in Context / Daniel Valle
010	Mythological Imagination, and the Flight from Capitalism / Hyonsob Kim
016	KRING
020	LIFE & POWER PRESS
024	Paris Olympic Memorial
028	Cheongshim Purification Center
032	Question of Making 'Being Unsangdong Architecture' / Jinho Park
036	2012 Yeosu EXPO Hyundai Motor Group Pavillion
040	House ONE : Chronotope Wall House
044	An Architect that Can Change / Seungman Baek
048	Gallery Yeh
052	Gallery The Hill
056	Urban Abstract Canvas: The Abstract Surface with Repetitive Partitions / Hayub Song
062	OCEANUS GROUP Haeundae Office
066	Kolong E + Green House
070	Primitive Boldness / Keehyun Ahn
076	Theater Contour (Namsadang)
080	Asian Culture Complex in Gwangju
084	Volume facade, Chessmen on chessboard / Sanghoon Youm
088	KTNG Culture Complex
092	Seongdong Cultural & Welfare Center
096	The Evolution of Experiential Skin Building / Jungwon Yoon
102	Hi Seoul Festival Stage Sculpture Palace pf May_Thousand Palace
106	White Cube Matrix: Paju Kindergarten
110	The Korean experimental architecture / Kyungsun Lee
114	A Thousand City Plateaus
118	Yeosu Expo 2012
122	Transformation of Compound Body to Social Imaginative Body / Youngbum Reigh
126	Shanghai Expo 2010
130	Mogyudowondo(Imagination City)
134	The Light and Shade of Skin Architecture / Eunseok Lee
138	Seoul Louis Vuitton Maison
142	Seoul architecture biennale pavilion
146	For the Universalization of Gi-un-saeng-dong / Hwang Yi
152	Jeonju Intangible Cultural Heritage Hall
156	SK Networks Gangnam Office
160	Shallow Sense, Deep Surface / Youchang Jeon
166	Dasan-Dong Fortress Wall of Seoul Parking and Cultural Center
170	Unbuilt Compound Body: The UnSangDong Experiment / Mannyoung Chung
174	Gallery Vogoze
178	Midong Electronics & Telecommunications Headquarter Office
182	Rare Architecture / Hangman Zo
186	White Quarter Circle
190	UGANDA Healing Mountain
194	Hetero-Pragmatics of Architecture / Helen Hejung Choi
200	Jeongok Prehistory Museum
204	Hannae Forest of Wisdom
208	Lie Sang Bong Tower

Daniel Valle (Department of Architecture, Chung Ang University)
다니엘 바예 (중앙대학교 건축학과)

운생동 건축사사무소는 그들의 시간을 패러다임으로 표현한다. 그들은 세계화와 오랜 기간에 걸친 민주주의 시대에서 처음으로 성장한 건축가 세대이다. 그들의 세대는 세계에서 가장 가난한 곳 중 하나였던 나라에서 가장 부유한 곳 중 하나로 변해가는 그들의 나라에 대한 급변적인 변화를 보아왔다. 이러한 변화는 진보에 대한 놀라운 집단적 욕망과 자본주의로 인해 매우 짧은 기간에 발생했다.

    수십 년에 걸친 극단적인 경제 개발의 부작용은 삶에 모든 면에 스며있는 성공에 대한 태도를 형성해왔다. 경제적 성취와 삶의 지위에 대한 집착은 유치원에서부터 시작된다. 경쟁적인 교육 방식은 성과와 결과에 초점을 맞추어 아주 어린 나이 때부터 성취에 대한 큰 압박감과 경쟁심을 부추긴다. 수십 년간의 극단적인 경제 개발의 부작용은 삶의 모든 문제에 스며드는 성공에 대한 태도 형성에 있다. 경제성과 삶의 지위에 대한 집착은 유치원에서 시작된다. 경쟁적인 교육은 젊은 사람들에게 성취에 대한 큰 압박감과 조잡한 경쟁심을 심어주어 아주 어린 나이부터 성과와 결과에 초점을 맞춘다. 사회에서의 차별성과 위치에 대한 강박적인 성향은 한국 도시의 경관에 고스란히 나타난다. 운생동 건축사사무소의 본거지인 서울시는 서로 옆에 놓여있는 별난

Unsandong Architects are a paradigmatic expression of their time. They belong to a generation of architects who has grown for the first time under a young democracy exposed to the forces of globalization and within an elongated period of peace. Their generation has seen an impressive transformation of their country passing from being one of the poorest places on earth to one of the wealthiest. This transformation occurred in a very short period of time driven by an incredible collective desire for progress and the fully adoption of capitalist's order.

    The side effects of decades of extreme economic development mentality have shaped an attitude towards success that permeates all matters of life. The obsession for economic achievement and status in life starts back in the kindergarten. A competitive education focuses on performance and results from a very early age giving young people great pressure for achievement and a crude sense of competitiveness. This translates into a reality where individualization and differentiation becomes critical for success and branding a tool to accomplish it. Everything became a product for the market including ourselves allowing social media giants to use our personal data for commercial purposes. Everything became "brand-able", including architecture. Homes are sold as products branded with simplistic slogans to seduce the public for a better life

건물들이 끝없이 이어지는 도시이다. 이런 건물들은 화합할 의도가 없고 오히려 각자가 주의를 끌려는 의도를 가고 있다. 서울은 복고풍의 VIP빌딩 또는"베데트(vedettes)"의 축적으로 인해 특이한 도시 조건을 갖추게 되었으며, 이는 더욱 독특하고 차별화된 건물의 수가 많이 존재 할 수록 그 다음 건물들도 그렇게 되는 결과를 낳았다.

      운생동 건축사사무소의 작업은 이러한 상황을 이용하여 건축물이 시장에 노출되도록 함으로써 오늘날 건축계에서 가장 인정받는 브랜드 중 하나가 되었다. 크링 금호 건물은 건축가가 건축주의 기업 가치를 효과적으로 반영해야 한다는 생각에 직면해 했던 대표적인 건물 중에 하나이다. 건물이 그 기업의 철학을 단순하게 표현할 수 있을까? 의사소통하기 위한 가장 좋은 전략은 무엇인가? 운생동 건축사사무소는 눈에 띄는 상징성과 은유를 사용하여 사람들의 시선이 외관으로 향하게 했다. 여기에는 건물이 상징적인 요소가 되어야 한다는 가정이 있다. 이 건물의 목표는 언론의 주목을 받고 결과적으로는 회사의 성공적인 제품이 되는 것이다. 역설적이게도 일단 건물이 상징화되면 디자인의 원래 중요성은 더 이상 중요하지 않게 된다. 찰스 젱크스(Charles Jenks)가 주장 하듯이, 현대의 상징적인 건물은 선진 사회에 대한 불신감을 반영하는 것이다. 상징적인 건물들은 기념물이 사라지면서 생기는 결과로, 특별한 의미가 첨부된 실체이다. 끊임없는 변화 하는 사회와 급격한 변화는 건물에 표현된다. 따라서 현대시대의 상징은 계산된 모호성으로 확산되고 변화 가능한 의미를 가질 필요가 있다. 크링 금호 건물 외관의 원은 궁극적으로 건물이 랜드마크가 된 순간에 자연의 조화에 대한 중요성과 언급을 잃게 됐다. 하지만, 이 새로운 형태의 현대 건축물은 어떤 것이든 더 나은 말이 될 수도 있다; 어떤 것이든 현대 사회에 대한 믿음의 약점의 결과로 상징적인 건물이 될 수 있다. 우리는 믿는 것을 멈추는 것이 아니라 오히려 모든 것을 믿으려 한다. 오늘날 마르셀 뒤샹의 '샘'이라는 작품처럼, 평범한 물체에 예술적 가치를 부여함으로써 거의 한 세기 전의 예술가들이 예상했던 대로 존경의 대상이 될 수 있다.

      이것은 운생동 건축사사무소가 상업적인 프로젝트라는 것에서 거리를 둘 때 더 영리하게 드러난다. 브랜딩의 제약에서 벗어나, 운생동 건축사사무소는 탄탄한 이론적 배경과 함께 창의성과 새로운 것의 중요성을 반영하는 흥미로운 작업들을 이어 나갔다. 그 배경은 일시성, 유행, 유동적인 사회, 디지털화, 복잡성 및 기타 현대적인 현상에 있다. "어울리기(Becoming)"라는 개념을 둘러싼 운생동 건축사사무소의 이론 구조는 끊임없는 변화의 상태에 그 기반을 두고 있다. 한국 사회의 순수한 본질은 항상 새로운 유행에 목말라 한다는 것이다. 건축가들은 이러한 변화무쌍한 상황이 평범함에서 특별함으로 옮겨갈 가능성을 가지고 있다고 믿는다.

while at the same time there is an extended perception that homes are a product to purchase for investment. This obsessive inclination to differentiation and positioning in the market translates clearly into the cityscape of Korean cities. The city of Seoul, home base of Unsangdong Architects, accounts for an endless catalogue of eccentric buildings seating next to each other. Each of them is trying to attract the attention of the surroundings with no intentions to collaborate but rather to attract. Seoul has been able to create a specific urban condition out of the accumulation of VIP buildings or "vedettes" in a sort of cycle that retro feeds, as more the number of eccentric buildings exist as more unique and differentiated the next one should be.

      The work of Unsangdong architects has taken this condition in their advantage accepting the potentials of exposing architecture to the market forces and by doing so they have become one of the most recognizable brands in today's architecture. The Kring Kumho Compound building is the most representative work among a group of buildings where the architects had to be confronted with the idea that a building should embed the corporate values of the client and, most importantly, communicate it effectively. Can a building express the corporation's philosophy bottled in a simplistic slogan? What is the best strategy to communicate it? Unsangdong Architects builds up clever semiotics and metaphors that drives the design towards a recognizable facade. There is an assumption that the building needs to become an iconic element. The building's purpose is to get the attention of the media and consequently became a successful product for the company. Paradoxically, the original significance behind the design is not important anymore once the building becomes an icon. As Charles Jenks argues, the contemporary iconic building is a reflection of misbelieve in developed societies. Iconic buildings arise as a consequence of the disappearance of the monument- an entity with a specific significance attached. The rapid transformation of society with constant changes and continuous arising of trends make the represented in a building obsolete in short time. Therefore, contemporary icons need to have a diffused and changeable significance with calculated ambiguity without a clear iconography. The rounded motives in the facade of the Kring Kumho Compound building ultimately lost their significance and references to Nature's harmony at the very moment the building became a landmark. This new type of contemporary architecture can become anything or better said; anything can become an iconic building as a consequence of the weaknesses of believes in contemporary societies. We not only stopped believing but rather we believe in everything. Nowadays anything can be object of veneration as anticipated by artists almost a century ago by giving artistic value to mundane objects -the R. Mutt Fountain i.e.

      It is when Unsangdong architects take distance from most commercial projects when their architectural proposition arises more cleverly. Away from the constrains of branding, Unsangdong Architects developed an

현대 사회를 형성하는 이러한 역동적 힘은 과거에 건축가들이 바라는 일반적인 주제였다. UN Studio의 다이어그램의 직접적인 공간화와 MVRDV의 Datascape에 따른 원시 데이터의 집중 사용으로 혁신적인 설계 프로세스를 만든 것은 작업의 복잡성을 통합시켰다. 운생동 건축사사무소는 사회의 역동성을 작업에 반영하기 위해 자체 방법론을 모색하고 있다. 이러한 아이디어는 건물의 입면에서 더욱 잘 드러난다. 미래주의(Futurism) 그리드 건물의 대각선 그리드는 서울시의 혹독한 환경으로부터 내부를 보호하는 외장장치일 뿐만 아니라 주변 환경과의 관계를 구축하는 방법도 아니다. 하지만 가장 중요한 것은 건물 내부에서 일어날 일을 우리에게 알려준다는 것이다. 돌출된 대각선은 일종의 구조적 불안정성을 시사하며, 이는 건물에 들어가기 전에 실제로 경계 또는 불안감을 유발하게 한다. 이 입면은 우리 시대의 허약함과 사업의 역동적인 본성에 대한 은유로 건물 안에서 일어나는 주요 활동들을 상기시킨다.

성동 문화복지관의 경우 전면의 대각선이 실내 공간의 노골적인 반영이다. 사실, 입면은 외피가 아니라 외부에 드러나는 부분이다. 이 건물은 건축가가 의사 소통하는데 있어 별 어려움 없이 역동성과 복잡성을 공고히 하는 건물이다. 놀라운 3 차원 공공 공간은 지상에서 5층으로 뻗어 있으며, 여러 개의 대각선 연결로 인해 내부와 외부로의 움직임이 스릴 있는 이미지를 만들어 낸다.

운생동 건축사사무소의 가장 주목할 만한 특징은 혁신에 대한 지속적인 태도일 것이다. 건축가는 여러 영역의 대지 요소를 통합하고 서로간의 충돌에서 균열이나 기회를 발견 할 수 있는 역량에 기댄다. 서울이라는 도시는 가능한 최상의 편의를 찾는 완전히 다른 유형의 건물들이 연속적으로 서로 나란히 놓여 있는 상반된 지도와 같다. 이러한 상반된 상황은 수십 년 동안 계속 되어 왔고, 이런 상황의 가능성을 잘 알아 본 것은 운생동 건축사사무소였다.

창조성에 대한 그들의 방법론적 접근은 OMA가 도입한 "사고"라는 개념과 닮았다. 네덜란드의 경우, 사고는 여러 층의 물질이 겹쳤을 때 일어나는 예상치 못한 충돌이었다. 운생동 건축사사무소에 있어 갈등 그 자체는 프로젝트의 원동력이다. OMA의 건축가들은 사고의 위치와 성격을 예측할 수 없었지만, 그들이 도시의 복잡성을 대표함으로써 일어나도록 했다. 이러한 사고는 가끔 나타나는 독특한 순간이었다. 운생동 건축사사무소의 경우, 갈등과 잠재력은 건물의 개념과 구성에 형태를 부여하는 설계 과정의 시작점이다. 파주 출판단지의 Press Pavilion에서 건물의 입면은 반대편의 체계적인 통합의 결과물이다. 건물의 모든 구성요소는 자체 DNA를 가진 원형으로 간주된다. 계단, 슬래브, 기둥은 서로 관련성이 있다. 그러기 위해서, 특정한 건축 요소의 유전적 부호는 늘 새로운 것으로 변형되어야 한다. 전통적인 중심

interesting array of work that continues reflecting the importance of creativity and newness with a solid theoretical background that constantly refers to the ephemeral, trends, fluid societies, digitalization, complexity and other contemporary phenomena.

Unsangdong Architects' theoretical constructs around the concept of "Becoming" is based on the status of constant change -a reflection of the pure essence of a Korean society always thirsty for new trends. The architects believe that this status of change possesses the potential to shift from ordinary to extraordinary.

These dynamic forces that shape contemporary society have been a common subject of desire for architects in the past. Both UN Studio's with their direct spatial formalization of the diagram or the intensive use of raw data by MVRDV's Datascape created innovative design processes to incorporate complexity into their body of work. Unsangdong Architects has also been in the search of their own methodology to incorporate society's dynamism into their work. These ideas are better expressed in the skin of their buildings. The Futurism Grid building's diagonal grid is not only the enclosure that protects the interior from Seoul's harsh environment nor the way the building establish a relationship with its surroundings but most importantly the facade prepare us for what is going to happen inside the building. The extruded diagonal lines suggest some kind of structural instability which in effect produces a state of alert or uneasiness prior to entering the building. The facade becomes a reminder of the fragility of our times as well as a metaphor of the dynamic nature of business -in fact the main activity taking place inside the building.

It is in the case of the SeongDong Cultural and Welfare Center where the diagonal lines in the facade become explicit reflections of the interior spaces. Indeed, the facade is not a facade but rather a section revealed to the exterior. It is in this building that the architects master the solidification of dynamics and complexity with an effortless needed to communicate it. An incredible three dimensional public space is stretched from ground level to the 5 floor with multiple diagonal connections that created a thrilling image of movement and buzz towards the interior and the exterior.

Perhaps the most remarkable and distinctive characteristic of Unsangdong Architects is their continuous attitude towards innovation. The architects rely in their capacity to integrate multiple layers of territorial elements and find the cracks or opportunities in the conflicts between opposites. A city like Seoul is indeed a continuous map of oppositions with completely different typologies of buildings lying next to each other finding the best accommodation possible. This opposition has been there for decades but it has been Unsangdong Architects who has better found the potentiality of this condition.

Their methodological approach to creativity resembles with the idea of "the accident" introduced by OMA. While in the case of the Dutch, accidents were

위치에서 건물 경계로 옮겨진 계단은 입면의 수직면과 함께 입면이 아닌 동시에 두 개의 층계가 되는 새로운 형태의 건축 요소를 만들어 내기 시작한다. 그것은 건물을 전시하고 경험하는 방식에 참신함을 가져다 주었다.

    운생동 건축사사무소의 작업에 접근 할 때 얻을 수 있는 첫 인상에도 불구하고, 그들의 건축물은 극도로 문맥적이다. 주변 상황에 대한 그들의 입장은 모방하거나 혼합하는 것이 아니라 궁극적으로 새로운 조건을 창출 할 여러 실체간에 대화를 부여하는 것이다. 이러한 논리는 아마도 그들이 건축가로서 성장한 도시의 본질에서 찾을 수 있을 것이다. 서울은 우리에게 상반된 것들이 공존할 수 있는 방법을 도시가 우리를 놀라게 하는 것과 같은 방식으로 보여주었고, 운생동 건축사사무소는 갈등에서 벗어나 혁신을 찾는 그들의 능력을 계속해서 찾아 나갈 것이다.

unexpected conflicts that occurred when various layer of matter where overlapped, for Unsangdong Architects the conflict itself is the motor of the project. OMA's architects were unable to predict the location and nature of the accident but they allowed them to happen as they represented the complexity of urbanity. These accidents were unique moments appearing occasionally. In the case of Unsangdong architects, the conflict and its potentialities are the starting point of the design process giving form to the idea and organization of the building. In Paju Book City's Press Pavilion the skin of the building is the result of a systematic process of integration between opposites. Every component of a building is considered as a prototype with its own DNA. Stairs, slabs, columns have all the potentiality to relate to each other. For that to happen, the genetic code of a particular architectural element must mutate into something new. The stairs displaced from the traditional core position to the perimeter of the building start mutating with the vertical plane of the facade creating a new type of architectural element that is neither a facade nor a stair but both at the same time. It is something that brings novelty to the way the building is presented and the way is experienced.

    Despite the first impression one can get when approaching Unsangdong Architects' work, their buildings are extremely contextual. Their position towards context is not of mimicking or blending but rather of imposing a dialogue between different entities that ultimately will create a new condition. The logic behind this position can be found, perhaps, in the essence of the city where they grew as architects. Seoul has shown us in many instances how opposites can coexist and the same way the city will never stop surprising us, Unsangdong Architects will continue developing their ability to find innovation out of conflict.

Hyonsob Kim (Department of Architecture, Korea University)
김현섭 (고려대학교 건축학과)

운생동의 건축에는 뭔가 특별한 게 있다. 우리의 시선을 사로잡는 기발한 디자인은 물론이거니와 그 이면에 흐르는 뭔가 특별한 논리가 있단 말이다. 그건 역시 운생동의 대표 건축가인 장윤규가 일관되게 펼쳐온 건축개념 혹은 이론에 기인한다고 할 수 있다. 이름하여 "복합체(Compound Body)". 2005년 출판된 단행본 『복합체』는 장윤규가 그동안 발전시켜온 건축이론의 집대성이다. 그에 따르면 "복합체"란 여러 요소들이 "완전하게 결합되기 전의 불안정한 상태를 나타내는 하나의 설정"이자 일종의 "가상의 텍스트"로서, 건축과 도시를 디자인함에 있어서 무수한 창조적 변형과 생성의 가능성을 내포한다.[1] "클립시티", "인간이 동물되기", "반응체", "스킨스케이프", "부유체", "인터랙티브 맵"이라는 일련의 표제어는 그런 복합체의 여러 단면들이라 하겠다.

우리는 이 같은 개념의 저변에 프랑스 철학자 질 들뢰즈의 사유가 짙게 배어 있음을 쉬이 발견할 수 있다. 게다가 장윤규도 자기 건축론에 그를 호출하는 것을 전혀 주저하지 않는다. 그러고 보니 언젠가 푸코는 20세기를 들뢰즈의 시대로 규정하기도 했었다. 그런 들뢰즈를 생각하면, 또한 현대 건축의 아방가르드적 사유에 실제로 이 프랑스 철학자가 보인 영향력을 생각하면, 장윤규의 경우가 뭐 그리 대단할까보냐. 허나 '말이 되는'

There is something special in the architecture of UnSangDong. There is a certain logic behind it, as well as an ingenious design that catches our attention. It is based on a consistent architectural concept or theory by Yoon Gyoo Jang, a representative architect of the UnSangDong. The book, Compound Body, published in 2005, is the culmination of the architectural theory that Yoon Gyoo Jang developed over the years. According to him, a "compound body" is "a set of elements that represent an unstable state before being fully combined," a kind of "virtual text," which implies the myriad creative transformations and possibilities of creation when designing architecture and cities.[1] "Clip City," "Becoming Animal," "Reactor," "Skinscape," "Fluid," and "Interactive Map" are examples of such compound bodies.

The underlying philosophy of the French philosopher Gilles Deleuze is deeply rooted in this concept. Yoon Gyoo Jang did not hesitate to reference Gilles Deleuze when describing his theory of architecture. Michel Foucault famously described the twentieth century as the age of Deleuze. Considering Deleuze and the influence this French philosopher had on the avant-garde idea of contemporary architecture, I wondered how innovative Yoon Gyoo Jang actually was. However, architects who consistently insists on a particular conceptual architectural theory are rare, so in this sense Yoon Gyoo Jangh was remarkable. There is a

건축론을 일관되게 펼쳐낸 건축가 워낙 희소한 우리네 실정에서 그이는 눈에 띈다. 비록 그의 말에도 허술한 논리가 발견되지만 말이다. 허허, 말이 되는 건축론이라 ... 현대 한국의 건축계에서는 1990년대 승효상의 "빈자의 미학"이나 아주 최근 제기된 황두진의 "무지개떡 건축" 정도를 꼽을 수 있지 않을까? 그런데 이들의 보수적이거나 현실적인 이론에 비하면 장윤규의 건축론은 훨씬 개념적이고, 실험적이고, 진보적이다. "개념적 건축을 실현하기 위한 실험적 그룹"이라는 운생동의 소개 글도 이를 잘 보여준다.[2] 더불어 그가 '갤러리정미소'를 열고 건축의 실험을 더 넓은 문화예술의 영역으로 확장해온 점 역시 주목할 바 아닌가.

신화적 상상력: "복합체"에서 "상상체"로

사실 운생동의 건축론인 "복합체"는 이제 더 이상 새로울 게 없는 듯하다. 우선은 『복합체』 출판 전후로 예화랑(2004~05)이나 Kring(2006~08)과 같은 운생동의 대표작이 국내외로 널리 소개되며 이 개념도 함께 알려지게 됐기 때문이다. 디자인으로 인한 건축개념의 동반 상승이다. 또한 더 직접적인 이유로는, "복합체" 개념이 그 후로 지금까지 훨씬 확장되거나 축약된 모양을 띠며 여러 차례 재출판되어 왔다는 사실을 들 수 있다. 첫 출판과의 차이라면 근래는 국제적 독자를 감안해 주로 영문본으로 책이 나온다는 점과[3] 매번 새로운 작품이 추가된다는 점이다.

하지만 이 출판물들 속에서 우리가 더 주목해 볼만한 변화는 2011년의 『Compound Body』가[4] 이전 책의 「인터랙티브 맵」 챕터를 「Mythological Imagination」, 즉 「신화적 상상력」으로 교체했고 이후의 작품집에서도 이 주제가 중요하게 다뤄진다는 점이다. 이는 "복합체" 개념의 다양한 측면을 새로이 묶어낸 일종의 결론과 같은 글로 볼만한데(『복합체』의 머리말에 언급한 오르페우스 신화와의 수미상관이랄까), "신화적 상상력"을 통해 현대 과학의 한계를 극복해야 함을 역설한다. 브레히트의 "거리두기" 혹은 "낯설게 하기"는 그 같은 상상력의 발현을 위한 매우 유효한 개념 틀이다. 거기 내포된 전통적 체제에 도전하는 아방가르드 정신이 "새로운 패러다임"을 가능케 한다는 것이다. 이 같은 주장은 매우 명쾌하며, 사실 아주 낯익다. 익히 알려졌듯 '예술'의 속성이란 게 기존의 고정관념을 깨는 낯선 상상력에 있으며, 이로써 지배적 패러다임을 넘어서는 비판의 가능성을 제시할 수 있기 때문이다. 그럼에도 운생동의 "신화적 상상력"의 논리가 우리에게 매우 도드라져 보이는 이유는, 앞서 암시했듯, 그간 우리네 건축계가 현실적 문제에 지나치게 구속될 수밖에 없었고, 또한 이처럼 과감하게 자신의 아젠다를 제시한 건축가가 적었던 데 기인한다. 흥미롭게도 그는 "신화적 상상력"을 논하며 "상상체(Mythological Body)"라는 개념을 새로이 끄집어낸다. ("Mythological Body"를 "신화체"가 아닌 "상상체"로 번역한 점은[5] "신화"와 "상상"을 등가로 봤기 때문 아닐까?) 신화적 상상을

fragile theory that can be found to support what he said. What is conceptual architecture theory? The "Beauty of Poverty" of Seunghyo Sang in the 1990's and the "Rainbow Cake Architecture" of Doojin Hwang are can be considered conceptual architecture in the field of contemporary Korean architecture. However, compared to their conservative, realistic theories, Yoon Gyoo Jang's theory of architecture is much more conceptual, experimental, and progressive. An introduction to the UnSangDong Architects Corporation, as "experimental group for realizing conceptual architecture," illustrates his architectural theory. [2] It is also noteworthy that he opened the Gallery Jungmiso and extended his architectural experiment to the field of wider culture and art.

Mythological imagination: From "compound body" to "imaginary body"

In fact, the compound body, the cornerstones of UnSangDong's architectural theory, s no longer seems new. Before and after the publication of the notion, the major works of UnSangDong, such as Yehwarang (2004~2005) and Kring (2006~2008), were widely introduced both domestically and abroad, and the concept of the compound body became commonly known. There was an accompanying increase in knowledge of the architectural concept, due to these designs. The book describing this concept has been reissued several times, both in expanded and abbreviated form. There are now differences from how it appeared in the first publication, because new works are added every time and an English version has been published for international readers. [3]

The changes we are most likely to notice in these publications include the "interactive map" chapter being retitled "mythological imagination" in the 2011 version. [4] This theme is still important in the subsequent versions. It deals with a conclusion of diverse aspects of the concept of the compound body. (It correlates with the Orpheus myths mentioned in the heading of the book.) It describes overcoming the limitations of contemporary science through a mythological imagination. This is also a valid conceptual framework for the manifestation of such imagination, as shown in the "alienation effect" or defamiliarization described by Brecht. The avant-garde spirit that challenges the traditional system embedded in it enables a new paradigm. This claim is clear and, in fact, familiar. As is well known, the property of art lies in the unfamiliar imagination that breaks existing stereotypes, and thus can present the possibility of criticism beyond the dominant paradigm. Nevertheless, the reason why the logic of the mythical imagination of UnSangDong seems to be distinctive for us is that, as suggested above, the field of architecture has been overly constrained by realistic problems. There are only a few architects who have boldly presented this agenda. Interestingly, Yoon Gyoo Jang discussed the mythical imagination and added the new concept of a "mythological body." The reason why he translated it as the "imaginary body" instead of describing it

현실화해 건축적으로 구축하겠다는 의지의 발로다. "상상체"는 "복합체"가 한 단계 진화한 개념 유형, 혹은 그것의 '알터 에고(Alter Ego)'인 셈이다.

그렇다면 "신화적 상상력" 및 "상상체"라는 개념의 정립이 운생동의 실제 디자인에 특별한 변화를 주는 계기가 됐을까? 글쎄다... 뭐 꼭 그렇다고 말하기는 어렵지 않겠나. 말과 디자인이 꼭 일대일 대응하는 것 만도 아니고, 이들이 이전 아이디어의 연장선상에 있는 개념쌍이기 때문이기도 하다. 게다가 아이디어의 생성과 소멸이 꼭 시간적 위계를 갖지도 않을 텐데, 들뢰즈적으로 말해 우리 생각의 그물망은 리좀적이라 할 수 있을 것이다. 장윤규가 수많은 텍스트의 관계망을 이미지화한 '하이퍼텍스트 맵'(2004)처럼 말이다. 그럼에도 불구하고 새로이 채택된 개념어는 운생동의 디자인을 조망케 하는 유용한 틀임에 틀림없다. 실제로 최근 출판된 운생동의 작품집들은[6] 「Mythological Imagination」이라는 소주제 하에 여러 작품을 아우른다. 그 가운데 2010 상해엑스포 기업관 응모작인 'Communi-Imagination'(2009)과 2012 여수엑스포 주제관 응모작인 'Ocean Imagination'(2009)이 대표적이다. 두 경우 모두 파빌리온 타이틀에 "Imagination"이 포함되기도 했는데, 아마 이 디자인 즈음이 "(신화적) 상상력"이라는 개념어를 발전시키던 때가 아닌가 싶다. "녹색 도시, 녹색 생활"이라는 기업관 주제에 대한 답인 'Communi-Imagination'은 실제의 자연과 인공의 자연을 결합시킨 "초현실적 소통(surrealistic communication)"을 의도했다. 파빌리온 전체는 소나무 숲을 담은 화분의 형상이며, 화분 자체는 여러 판을 적층시켜 우리나라의 자연지형을 본뜬 모양새다. 건물 사면에 새겨진 사계절 풍경의 실루엣은 디지털 스킨과 함께 환상적 분위기를 연출하는데, 옥상 소나무 숲의 호수에서 떨어지는 물의 커튼은 그 같은 분위기를 배가시킨다. 내부 전시공간이라든지 에코시스템 등의 실무적 이슈는 논외로 하자. 운생동의 이 파빌리온 계획안은 참신한 상상력을 바탕으로 자연과 테크놀로지의 결합을 꿈꾼 "복합체"이자 "상상체"라 하겠다.

마찬가지로 'Ocean Imagination'도 생태적 관점에서 "녹색(자연)과 상상력"을 첨단기술로 조화시킨다는 아이디어를 내세우는데, 기본 골자는 앞의 사례와 거의 유사하다. 가장 큰 차이라면 마린 테크놀로지 등 바다와 관련된 주제를 포함한다는 점과 독특한 건물 형태가 강렬한 힘을 선사한다는 점에 있다. 이 파빌리온은 커다란 원을 프레임으로 한, 내부가 불규칙하게 파여진 바퀴와 같은 모양으로서 바다 위에 둥둥 떠서 빙그르르 돌아갈 것만 같다. 원의 프레임에 지지된 각 공간은 숲과 같은 초목의 공간 및 수족관과 같은 해양 관련 공간으로 대별할 만한데, 각각의 외피의 얼개는 마치 나뭇잎의 잎맥 및 파도의 물거품을 은유한 듯하다. 그리고 파빌리온 안쪽의 커다란 보이드는 미디어 아트와

as mythical may be because he believed myth to be equivalent to the imagination.[5] He was willing to build architecture by realizing the mythical imagination. The imaginary body is the compound body advanced one step further; it is its alter ego.

Did the concepts of mythical imagination and imaginary body make special changes to the actual designs of UnSangDong? This is difficult to say. The words don't always correspond to the design, but they are also concepts that serve as extensions of previous ideas. Moreover, the creation and extinction of ideas do not necessarily follow a temporal hierarchy. In the words of Deleuze, the net of our thoughts can be called rhizomatic. It is like the "Hyper Text Map" (2004), where Yoon Yoon Gyoo Jang imagined a network of numerous texts. Nevertheless, the newly adopted concept is a useful framework for visualizing UnSangDong designs. The recently published works of UnSangDong cover various creations under the heading of the mythological imagination.[6] Among them, "Communi-Imagination (2009)," which was the entry for the 2010 Shanghai Expo, and "Ocean Imagination (2009)," which was the entry for the 2012 Yeosu Expo, were representative. In both cases, the pavilion entitled "Imagination" was included. This was probably when the mythological imagination concept was developed. The theme of the corporation pavilion, "Green City, Green Life," corresponded to "Communi-Imagination," which was intended to be a form of "surrealistic communication" that combined real nature with the artificial. The entire pavilion was in the form of a flowerpot containing pine forests. The flowerpot itself was a pattern of laminated plates, simulating the natural topography of Korea. The silhouette of the four seasons carved on the faces of the building created a fantastic atmosphere with digital skins, and the curtain of water falling from the lake of the pine forest, which was on the roof, doubled the atmosphere. We will not discuss practical issues such as internal exhibition spaces or eco-systems. UnSangDong's pavilion design was a compound body and imaginary body that combined nature and technology, based on a novel imagination.

Likewise, "Ocean Imagination" also introduces the idea of a harmony between "green (nature) and imagination," from an ecological view. The main concept is similar to that of previous examples. The biggest difference is that it includes sea-related topics such as marine technology, and the distinctive form of the building offers intense strength. The pavilion is like a wheel with an irregularly shaped inner part based on a frame with a large circle; it seems to float on the sea. Each space supported by a circle frame is divided into a vegetation space such as a forest, and a marine related space like an aquarium. The framework of each outer skin is similar to leaf veins and sprays of waves. Because of the large void inside, the pavilion was designed to serve as a space for media arts and laser shows. If this had been realized, it would have been possible to experience yet another type of imagination. Although UnSangDong's design is based on experimental imagination, these two

레이저 쇼 등을 위한 공간으로 마련된 까닭에, 실현됐다면 또 다른 상상력의 체험이 가능했을 것이다. 운생동의 디자인이 모두 실험적 상상력을 기반으로 한다지만 이 두 건물에 상상력이 특히 더 요구됐을 법한데, 아무래도 엑스포 파빌리온이 상징성을 훨씬 중시하는 임시 시설물이라는 사실 때문이었음이라. 실용성에 목적을 두며 반영구적으로 유지관리를 해야 하는 일반적 건물과는 달리 현실과의 타협이 상대적으로 덜 했을테니 말이다.

그럼에도 불구하고 운생동은 상상을 현실화하기 위해 부단히 노력해왔으며, 지금까지 실현된 다수의 작품을 통해 상당한 성취를 거뒀다고 할 수 있다. 최근 실현된 디자인 중에는 몽유도원도 이상봉타워(2012~18)를 흥미로운 사례로 거론할 만하다. 이 건물은 몽유도원도의 산세일지 구름일지, 꿈속의 몽롱한 풍경을 추상화해 건물 입면에 적용한 점이 무척 인상적이다. 여러 판이 겹쳐져 울룩불룩한 파사드를 이루는데, 굳이 비교하자면 그 판들의 적층은 'Communi-Imagination'의 유형을 수직화한 셈이다. 운생동은 그 입면의 테마만을 가지고 독립적인 설치 조각을 만들어 2015년 베니스비엔날레에 전시하기도 했다. 조선시대 안견이 그린 '몽유도원도' 자체가 이 조각의 이름으로 차용됐는데, 동양적 유토피아를 "상상의 도시"로 형상화한 결과다. 이는 데카르트의 명석판명한 논리구조에 들뢰즈적 변형과 해체를 가하고자 하는 운생동의 노선을 반영한 것에 다름 아니다. "개념적" 건축이기도 하고 도시이기도 한 이 '몽유도원도'를 운생동은 "신화적 상상력"의 주요 사례로 간주하고 있다.⁷

"브랜드스페이스", 그리고 자본주의로부터의 탈주
헌데, "복합체"든 "상상체"든, 장윤규와 운생동의 아이디어에서 짐짓 재고해야 할 바가 하나 있다. 그들이 들뢰즈적 어법을 적극 내세워서 말인데, 혹시 그들의 디자인이 자본주의 시장의 마케팅 전략을 고스란히 긍정함으로써 들뢰즈의 기본적 방향성에 역행하는 것은 아닐까 하는 문제다. 이런 혐의는 운생동의 대표작인 금호복합문화공간 Kring을 설명한 글 "브랜드스페이스 되기(Being a Brandspace)"에 특히 깔려있다.⁸ 여기서 장윤규는 기업을 위한 건축공간이 해당 기업의 이윤창출만이 아닌 더 넓은 공공적이고 문화적인 파급효과를 가져올 수 있다고 주장함으로써 기존의 관례로부터 상당한 진보를 보인 게 사실이다. 그럼에도 불구하고 "공간"이든 "문화"든 모두 가감 없이 "마케팅"과 직결시키는 논리는 "상품으로서의 건축", 즉 건축의 상품화에 적극 복무함으로써 자본주의의 욕망에 그대로 투항하는 게 아닌가 싶다. 자본주의에 대한 들뢰즈의 비판적 입장이 설령 야누스의 얼굴처럼 보일 때도 있지만, 혹자의 말처럼 들뢰즈야말로 "놀라울 만큼 강하고 교활한" 자본주의에 대한 비판을 "가장 멀리까지 끌고 간 사상가"다. 끝없는 욕망에 기초한 자본주의는 그로 인한 위기를

buildings must have required so much more imagination. The pavilion at Expo was a temporary facility that placed more emphasis on symbolism. Unlike general buildings that have to be semi-permanently maintained for practical purposes, in temporary structures there is less compromise with reality.

UnSangDong endeavored to realize the imagination and thus far has accomplished considerable achievements through numerous works. Among the recently realized designs is Mongyudowondo, Lie Sangbong Tower (2012~2018). This building is impressive in that it is abstracted from the sun, cloud, and dreamy scenery on Mongyudowondo, which are applied to the facade of the building. Several plates overlap to form an uneven facade, meaning that the lamination of the plates is a verticalization of the "Communi-Imagination" type. UnSangDong created an independent installation sculpture with only the theme of the facade, the Lie Sangbong Tower, and exhibited it at the Venice Biennale in 2016. The name of this work was borrowed from the, "Mongyudowondo" masterpiece by Kyun An, completed during the Chosun Dynasty. This art is a symbol of the Asian utopia, "the city of imagination." This reflects UnSangDong's goal, which attempts to add transformation and demolition through Deleuze to Descartes's clear logical structure. UnSangDong believes that "Mongyudowondo," which is both a conceptual structure and the city itself, can also be considered the main example of the mythical Imagination.⁷

"Brandspace," and an escape from capitalism
However, whether is a compound body or imaginary body, there is one thing to reconsider from the ideas from Yoon Gyoo Jang and UnSangDong. As they often quoted Deleuze's words, it is doubtful that their designs would be contrary to the basic direction of Deleuze firmly affirming the marketing strategy of the capitalist market. This question lies particularly in the article "Being a Brandspace," which describes Kring, a Kumho culture complex that is representative UnSangDong's work.⁸ In this article, Yoon Gyoo Jang made considerable progress from existing practice by insisting that the space for the enterprise could bring about a wider public and cultural ripple effect, and not just create profit for the company. Nevertheless, the logic of direct connections with space or culture to marketing seems reflected in the desire for capitalism expressed by actively commercializing "architecture as a commodity." While Deleuze's critical position on capitalism may look like Janus's face, "Deleuze is a thinker who has dragged the criticism of incredibly strong and cunning capitalism further." Capitalism is based on endless desire, and thus is bound to contain the crisis. The possibility of getting away from capitalism is Deleuze's escape. The reality is never easy. Architecture seems to have no willingness to escape from its original capital. The reality of architecture is that it cannot but be forced into reterritorialization with a huge mechanical system that swallows everything, even if we try to deterritorialize it from capital.

내포할 수밖에 없는데, 그로부터 벗어날 가능성을 엿보인 것이 바로 들뢰즈의 탈주라는 것이다. 물론 현실은 결코 녹록치 않다. 건축이란 본래 자본으로부터 벗어날 여지가 전혀 없어 보이니 말이다. 자본으로부터의 "탈영토화"를 아무리 감행해도 모든 것을 집어삼켜버리는 그 거대한 기계 시스템으로 "재영토화"될 수밖에 없는 건축의 현실... 그래도 말이다, 개념적이고 실험적인 건축을 주창하는 운생동이니, 그리고 상당한 가능성을 보여줬으니, 한 발짝만 더 과감한 탈주를 시도한다면 어떨까? 작금의 세계 자본주의가 파놓은 홈 파인 공간을 가로질러 유유히 탈주하며, 좀 더 매끈한 공간으로 유희해보잔 말이다. 현실 너머 상상으로 기운생동! 이것이 말 되는 건축론을 펼친 운생동에서 한국 건축계가 욕심내 봄 직한 요청 아닐까?

Since UnSangDong insists on conceptual and experimental architectures and has shown us numerous possibilities, why not try one more bold breakout? Why not escape unhurriedly from the gap made by world capitalism and create a smoother space? Beyond reality, and on to the imagination! Is it possible that Korean architecture can rely on UnSangDong to make it happen?

1 장윤규, 『복합체』, 간향미디어랩, 2005, p. 20.
2 운생동 웹사이트 및 여러 출판물.
3 운생동이 2006년과 2007년 미국의 『Architectural Record』와 영국의 『Architectural Review』에서 조명 받았던 일은 국제적 독자를 더욱 의식케 한 계기가 됐을 것이다. 한편, 확장 혹은 축약된 『복합체』 영문본으로는 『Compound Body』(UnSangDong, 2011), 『Compound Body』(l'ARCA, 2017), 『UnSangDong Architects』(UnSangDong, 2018) 등이 있다.
4 운생동이 자체 출판부(UnSangDong Publication)에서 발간한 이 책은 『복합체』 이후 수행된 작품까지 모두 아우르는, 훨씬 크고 두껍고 럭셔리한 영문판 포트폴리오다. 텍스트가 보강됐고, 작품집으로서의 성격이 더 강조됐으며, 장윤규 개인을 넘어 신창훈을 포함한 조직 "운생동"을 내세운다.
5 혹은 거꾸로 "상상체"를 "Mythological Body"로 번역했다고 할까. 이것이 장윤규의 의도인지 (번역자의) 실수인지도 사실 불확실하다. 『Mythological Imagination』의 한국어 텍스트는 『한국건축의 새로운 지평』(UnSangDong, 2011) 참조.
6 『Compound Body』(2017)와 『UnSangDong Architects』(2018) 참조.
7 『UnSangDong Architects』, UnSangDong, 2018, pp. 201-208.
8 이 개념은 『Compound Body』(2011) 이래의 문헌에 계속 나오지만 때로 "Brandspace"와 "Brandscape"를 혼동해 쓰기도 하는데, "Being"과 "Becoming"의 구분도 좀 모호하다. 한국어 텍스트는 『한국건축의 새로운 지평』(2011) 참조.

1 Yoon Gyoo Jang, "Compound Body" Ganhyang Media Lab, 2005, p. 20.
2 UnSangDong website and other publications
3 The works of Unsangdong were introduced and highlighted in the Architectural Record, USA, in 2006, and the Architectural Review, England, in 2008, attracting international readers. Also, the expanded and abbreviated English versions of Compound Body were translated with titles as follows: Compound Body(UnSangDong, 2011), Compound Body(l'ARCA, 2017), UnSangDong Architects(UnSangDong, 2018).
4 The book Compound Body by UnSangDong Publications is a much larger, thicker, and more luxurious English-language portfolio that encompasses everything that has been done since the book was first published. The text has been reinforced, its workbook-like characteristics have been emphasized, and the organization, "UnSangDong," including Yoon Gyoo Jang and Changhoon Shin, have become the focus.
5 Or, conversely, "imaginary body" can be translated as "mythological body." It is uncertain whether this was intended by Yoon Gyoo Jang or a mistake. Please refer to "New Horizon in Korean Architecture" (UnSang Dong, 2011) for the Korean language version of Mythological Imagination.
6 Refer to Compound Body(2017) and UnSangDong Architects(2018)
7 『UnSangDong Architects』, UnSangDong, 2018, pp. 201-208.
8 This concept has been addressed regularly in the literature since Compound Body(2011), but is sometimes confused with "Brandspace" or "Brandscape." The distinction between "being" and "ecoming" is also vague. Please refer to "New Horizon in Korean Architecture" (UnSang Dong, 2011) for the Korean language.

KRING (KumHo Culture Complex)
크링 (금호복합문화공간)

Location: 9968-3 Daechi-Dong, Gangnam-Gu, Seoul, Republic of Korea **Use:** Temporary Building **Building Scope:** 3F **Site Area:** 4,110.9m² **Building Area:** 3,153.58m² **Total Floor Area:** 7,144.53m² **Building Coverage Ratio:** 76.71% **Floor Area Ratio:** 173.06 % **Structure:** Steel Structure+Stainless Steel, Sandwich Panel **Principals in Charge:** Jang Yoon Gyoo, Shin Chang Hoon, Kim kyeung Tae **Design Team:** Kim Sung Min, Moon Sang Ho, Kim Se JIn, Kang Seung Hyeun, Kim Bong Kyun, Goh Young Dong, Yi Na Ra **Client:** KumHo E&C **Photographer:** Jaekyeong Kim(Model), Sergio Pirrone

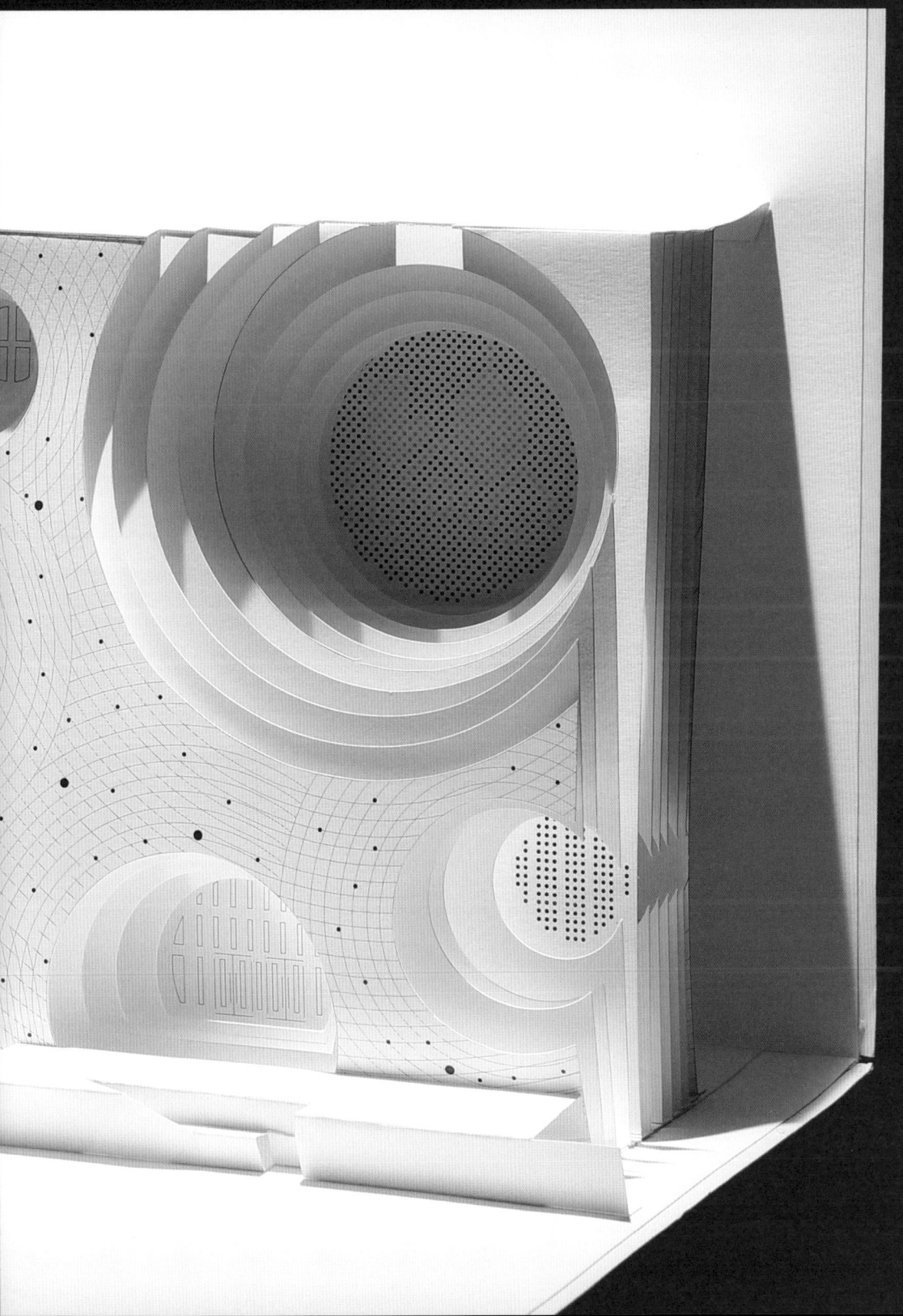

사람들의 꿈을 대변하는 건축공간을 구성할 필요가 있다. 사회적 정체성은 생산하는 것보다 소비하는 것과 관련되어 있다. 일반인들이 예술적인 작품을 체험하게 하는 방식을 '상품화된 공간'을 제공하는 형식으로 채용할 수 있다. 소비자의 욕구를 자극하는 현상으로 건축을 드러나게 하는 것이 그 예의 하나라 볼 수 있다. 이제 건축가들도 하나의 브랜드적 가치에 기대며 상품을 만들어 내는 것처럼 건축을 실행할 필요가 있다. 상품을 만들어내는 방식과 상품을 파는 근본적인 방식에 대한 이해가 요구된다. 디자인을 판다는 것은 단순히 소비자의 요구에만 반응하는 것이 아니라 감성을 자극하고 다음 요구를 원하게 하는 상황을 연출하는 데 있다. 이러한 상황을 만들어내고 리딩하는 도구로서 '디자인 마케팅적 사고'가 작용한다. '금호 문화관'이라는 기업적인 성향의 클라이언트를 통하여 '브랜드적 접근'을 통한 프로그램적 변화를 시도 하였다. 외부적으로 기업의 문화를 적극적으로 홍보 할 수 있는 방법과 기능적으로 주택 전시라는 새로운 프로그램과 공간적 가능성을 획득하는 브랜드 스페이스를 제안한다. 더욱더 닥쳐올 정보화된 사회에 건축가는 새로운 브랜드의 가치로 무장하여야 한다.

We, Unsangdong Architects, are proposing a rather unfamiliar approach to architecture, <Making a brandspace>, which was exemplified in Kring_Kumho Compound Culture Space. Brand strategy, once used to be employed only by profit organizations as a means to create monetary assets, has evolved endlessly to reborn. Profit organizations are under more severe surveillance than ever before by public and required to demonstrate social responsibility of their own brand. Branding had been a useful tool for marketing, now it is re-categorized under brand new realm, which includes spatial concept. We would like to focus on new models of branding such as <Concept of the third space>, or <Spatial Design Marketing>. <Spatial Design Marketing> defines any and all marketing activities by profit organizations to provide a space as a direct means to satisfy customers' needs or desire, or offer a spatial experience to customers by making themselves reconsider a profit organization or the image of that particular brand with space. Space, as well as spatial design, reinforces its image as a medium for facilitating communications between companies and consumers, as opposed to simply perform its previous role to provide a place for containing customer's human behavior.

LIFE & POWER PRESS
파주 생능출판사

Location: 507-12 Munbal-li, Kyoha-eup, Paju City, Republic of Korea **Use:** Office **Building Scope:** B1F ~ 7F **Site Area:** 727.20m^2 **Building Area:** 352.50m^2 **Gross Floor Area:** 995.72m^2 **Building Coverage Ratio:** 48.47% **Gross Floor Ratio:** 135.24% **Structure:** R.C **Principals in Charge:** Jang Yoon Gyoo, Shin Chang Hoon **Client:** Kim Seung Ki **Photographer:** Namgoong Sun

CULTURAL TOPOGRAPHY
- STACKING CONTOUR

Cultural Topography_충돌의 지형

우리가 제안하는 지형은 갈등과 충돌을 야기하는 양립할 수 없는 부분을 건드리는 사이의 공간이기도 하다. '새로운 지형의 그릇'을 제안하면서 작가들과의 충돌을 피할 수는 없다. 그릇과 내용물, 형식과 내용, 전체와 부분의 관계에 충돌이 생기는 것은 분명한 문제였다. 우리는 문화간의 충돌을 회피할 수 없다. 서로간의 이해와 협력에 의해서 새로운 인터랙티브의 구조를 실현하여야 하며, 서로의 영향을 주고받는 거대한 맵의 체계를 구성하여야 한다. 서로의 이야기와 토론을 즐기는 '충돌의 지형'을 탐구한다. 생능출판사라는 프로젝트를 통하여 보르헤스적인 가상의 텍스트로 작용하는 '새로운 지형의 그릇'을 상상한다. '새로운 지형의 그릇'은 균질화된 전시공간에 주름진 지형의 공간을 삽입하여 창조적 공간의 체험을 제공하는 데 있다. 우리가 설정하는 '지형'은 일반화된 형식에서 탈피하여 체험과 공간인식의 다른 차원을 이동하는 내용과 방식을 담는 틀이 된다. 변형되는 지형은 물리적인 형태만을 담고 있는 것이 아니라 문화적인 컨텐츠를 공유하는 관념적인 지형의 일부이기도 하다. 이것은 문화지형과 같은 구도를 가진다. 생능출판사를 통하여 자연의 대지 형상을 추상적으로 변화시킨 '쌓아 놓은 듯한 콘타 지형도_Map of Stacking Contour'를 제안한다. 단면의 변화를 이용한 보이드와 솔리드 공간이 교차하는 지형의 쌓기에 의해서 구성되는 공간을 제안한다. 지형의 충돌에 의해 파생된 계단과 같은 쌓기는 피라네지(piranesi)의 공간적 구조와 맞물려 있다. 사무실이라는 프로그램적 한계를 외피공간의 확장에 의해 삽입되는 새로운 가능성의 프로그램인 시저긴 휴식이 가능한 계단형의 발코니와 교차시킨다. 삽입된 프로그램에 의해서 휴게, 퍼포먼스, 갤러리, 세미나 등의 다양한 프로그램의 변화를 예측하게 한다.

Life & Power Press Cultural topography - Stacking Contour

We know that we exist not as an individual, but as a unity of many, or even numerous individuals connected to each other. Creating an Publisher's space is not just setting up the exhibition and office space and putting books in it, but drawing a map that covers all the works in the venue. It depends on the map how the viewer will appreciate the space and books. Through the <publisher of Life & Power>, I have sought to insert 'the vessel of new topography' into homogeneous space. 'The vessel of new topography.' an intervention of a new imagination into the into the space, works as imaginary texts like those of Jorge Luis Borges. We will make-the "Contour Topography" a space for the' new experience. The Contour space will reflect the real topography by applying the abstract image of it to the floor. The theatrical "Topography" works as canvas for the life, but changes through various new media include book. The multi-purpose topography is a interactive map of office, club, show, seminar, lecture, performance, etc.

Paris Olympic Memorial
파리 올림픽 메모리얼

Location: Paris, France **Use:** Landmark + Gallery **Structure:** Reinforced plastic cell **Exterior finishing:** Plastic wall, Full-color LEDs **Structure:** Reinforced plastic cell **Photographer:** Unsangdong(Model)

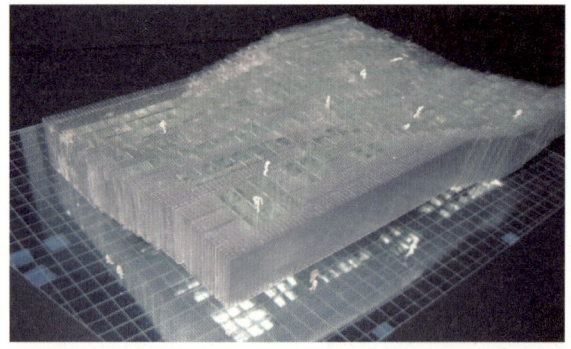

추상화된 지구의 맵을 통해서 올림픽 랜드마크를 제안한다. 전세계 종족과 지역을 추상화하여 셀을 만들고, 이들을 조합하여 지도로 표현한다. 셀은 수직적으로 돌출된 형식으로 전체의 랜드마크 맵을 구성한다. 셀의 수직적인 면을 퍼즐의 조합에 의해서 긴결하게 붙여나가며 입체적이고 무중력적인 매트를 구성한다. 기존의 대지로부터 새롭게 놓인 셀들의 맵은 입체화된 매트와 같은 공간을 이루며 기존 대지와의 소통과 공간적 경험, 외부적 연속과 내부적 네비게이션을 가능케하는 랜드마크적 연결고리를 제공한다. 셀들의 집합과 새로운 무중력적인 배열에 의해서 풍경의 픽셀, 조경의 픽셀, 빛의 픽셀, 색채의 픽셀, 유리의 픽셀 등으로 변화되어 살아있는 랜드마크가 된다.

We propose an olympic landmark using an abstract map of the earth. Every race and region in the world is represented by a map of abstract cells. The whole landmark map is composed of vertically extruded cells. 3-dimensional and zero gravity matrix is constructed by mechanical puzzle joints which tightly connect vertical sides of each cell. The map of newly positioned cells from the existing ground composes the 3-dimensional matt-like space and provides connection link to communicate with the ground, to experience the space, and to navigate inside and outside the landmark. Through a re-organizational assembly of anti-gravity cells, it becomes a live landmark, which can be transformed into pixels of scenery, pixels of landscape, pixels of light, pixels of color, pixels of glass, and etc.

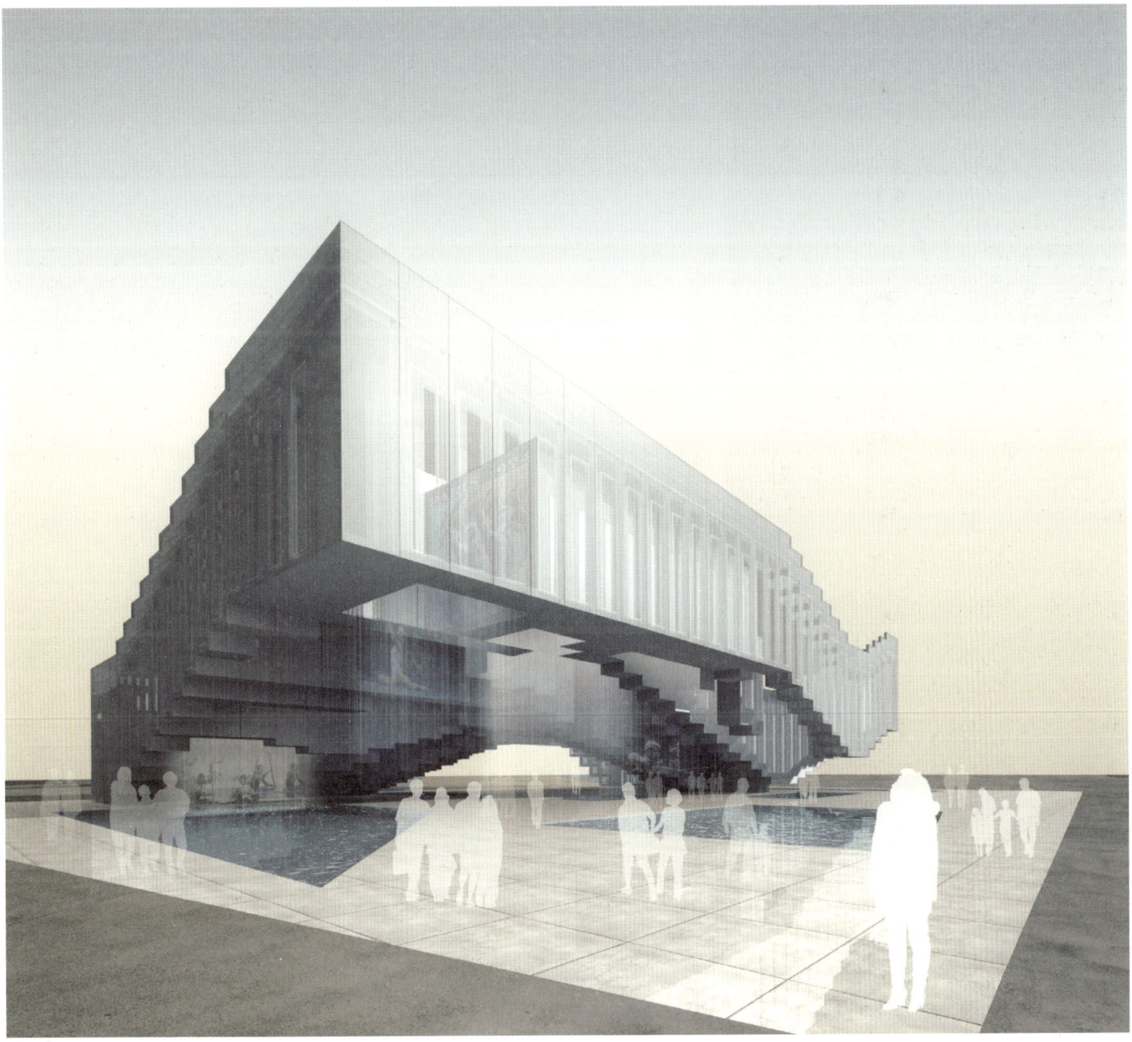

Cheongshim Purification Center
청심 물문화관

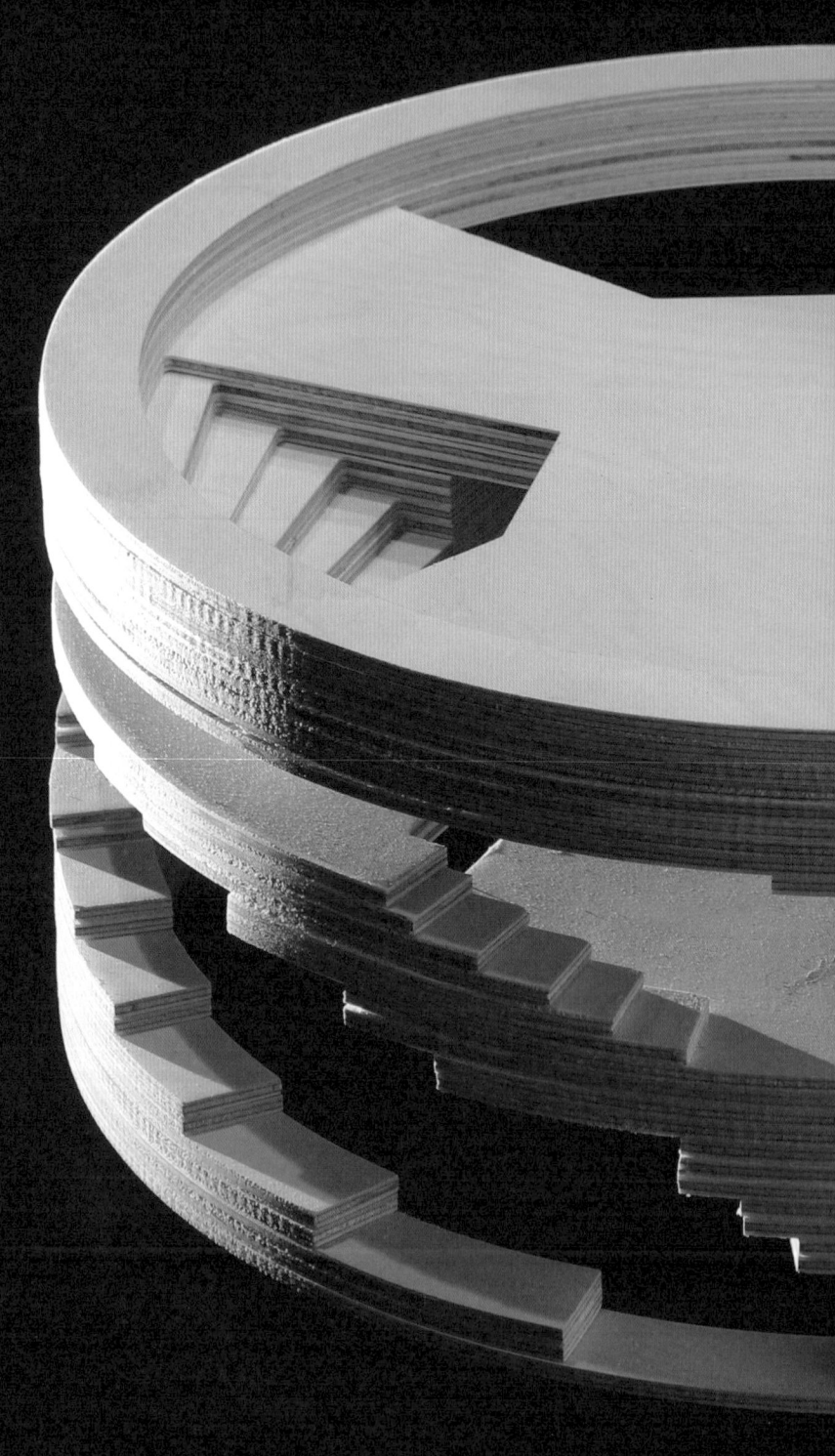

Location: 555-9 SongSan-li Seolak-myeon Gapyung Kyeunggi-do, Republic of Korea **Use:** Purification Center
Building area: 505.00 m² **Gross Floor Area:** 637.40 m² **Building Scope:** B1 ~ 2F **Structure:** R.C. **Photographer:** Jaekyeong Kim(Model), Fernando Guerra, Sergio Pirrone

'Water Circle'은 청심단지개발을 통해서 생성되는 다양한 오수를 정화하기 위해서 만들어야만 하는 인프라시설인 오수정화시설을 새로운 건축적 체험과 정신적 정화로 변환시키는 흥미로운 프로젝트이다. 일반적으로 기존의 오수정화시설은 오염된 물을 정화하여 기능만을 수행하는 것을 목적으로 하고 있기 때문에 환경과 미적인 고려가 전혀 없는 혐오시설로 만들어 졌다. 'Water Circle'은 이 혐오시설을 미술관적 체험과 결합시켜 실제로 일어나는 물의 정화의 과정을 인식하고 동시에 물에 관련된 다양한 체험을 공유하는 '환경 인프라와 물의 교육'이라는 사회적 장치로의 변환을 제공하는 것이다. 즉 실제로 작동되며 주변에 오염된 물을 정화하는 장치가 됨과 동시에 물의 소중함을 동시에 느끼는 사회적이며 정신적인 정화를 이루어내는 건축을 생성하는 것이다. 단순한 오수처리장의 기능을 넘어선 교육, 전시, 환경설비공간, 자연공간등의 복합적 기능의 연계를 통해 자연의 생명력을 담는 창조적인 인프라가 되는 것이다.

Most of all, we focused on a robust, simple and tranquil architecture intuitively as we take up the form and space. Another point we considered is a vigorous architecture which evoke water vitality in still emotion. The circular form symbolises pure crystal of living organisms. Circular plan which is 32 metres in diameter, 11 metres in height soars from the ground. Firm, still and refined curve wall opens with gradual, arbitrary and rhythmic rules. We expect another tone colour through the open and solid curved wall. Musical facade is achieved through this project it plays in andante, moderato and allegro. It conveys silence and movement simultaneously. The flux of water and changing biological image has always been interesting architectural either subject and material. The concrete outer wall stacked on entrance freely metaphorically indicate to a wave, and it provide to experience the between space of outer and inner space. This space holds the curiosity of the space stepping into the water when we arrive in the shore. The space beneath water bears fluidity that is slowly sucked in the space of condensed crystallization at the moment.

Question of Making 'Being Unsangdong Architecture'

운생동스러운 건축 만들기에 대한 물음, Unsangdong - Resque

Jinho Park (Department of Architecture, Inha University)
박진호 (인하대학교 건축학과)

현대 한국 건축을 둘러싼 모종의 강박 중 하나는 뭔가 새로운 것을 만들어야 한다는, 혹은 뭔가 다름을 추구해야 한다는 건축가들 스스로의 압박이다. 급격한 산업화로 인해 양산되는 평범한 건축으로 가득 찬 도시의 지루함도 그렇거니와, 상업성이 난무하는 무의미한 건축이 도시의 환경을 이루어가는 것에 대한 강한 저항의식도 서려 있다. 이러한 의식은 결국 진부함으로부터 벗어나 건축가들의 개별의식이 강조된 건축을 향한 길 찾기로도 이해될 수 있다.

척박한 건축환경 속에서 작업하는 한국의 건축가들은 새로운 건축을 향한 열망과 현실 사이의 모순에 절망하기도 한다. 시대정신이나 의식의 결여를 비웃기라도 하듯, 여전히 견고한 근대적 질서라는 영향력 아래 유지되는 현실적 벽에 좌절하기도 한다. 관습을 지키려는 혹은 관습을 깨려는 문제는 서로 항상 충돌하며 타협하며 합의점을 찾기도 조용히 잊혀지기도 한다. 창의적 건축의 발현을 위한 일상에서의 노력은 욕망과 절망이 반복되는 상황의 지속이다.

탈중심 혹은 탈이념의 시대의 한국 현대건축에는 개별적이고 독립된 방식의 목소리가 생성되기도 하다가 사라지기도 하기 때문에 한 시대를 표상하는 보편적인 공감대나

One of the particular compulsions about modern Korean architecture is the pressure that architects put on themselves to create something entirely new or different. There is a strong sense of resistance against the boring cityscape that consists of ordinary buildings that are a result of the rapid industrialization, and against the rampant construction of futile commercial buildings making up the city environment. This consciousness can also be understood as a path to architectural emphasis on the individual consciousness of the architects, which eventually escapes the stereotype.

Korean architects, who work in a harsh architectural environment, at times despair for the contradiction between reality and the desire for new architecture. As if laughing at the lack of consciousness or a spirit of the times, they are at times frustrated by the realistic walls that remain standing under the influence of the firm modern order. The problem of keeping or breaking customs always conflicts with the other, or makes compromises, or finds common ground, or is even quietly forgotten. Efforts made in everyday life for the emergence of creative architecture are a continuation of a situation with desire and despair on repeat.

In Korean modern architecture in the era of decentralization or de-ideology, voices of individual and

변화의 탄력(momentum)을 형성하기에는 그 에너지가 미약하다. 허나 고정관념의 울타리 속에 변화를 시도하지 않고 일상성과 평범함의 틀 속에 스스로를 가둔다면 그들의 목소리는 사회 전반에 증폭되지 못하고 소통도 되지 않음으로써 고립을 자초하게 되고 결국 발전을 저해하게 된다.

이러한 상황에서도 새로운 건축 만들기의 변화를 실천해 가는 작지만 다양한 목소리들이 한국의 현대건축에 존재한다. 일상의 코드화된 관념에서 벗어나 건축가의 개별적 건축문법과 담론으로 그 탈출구를 만들어 가고 있다. 탈이념적이며 탈근대적 어휘로 중심성이 흐릿한 지금의 건축에서 오랫동안 우리 건축을 지배하고 있던 규칙적이며 반복적인 틀에서 탈피하고자 하는 노력이다.

그 흐름에 운생동이 있다. 지난 수 년간 운생동에서 해온 작품들을 보면, 일상의 작품활동을 견인해 가는 와중에 견고한 근대적 틀을 조금씩 지워가는 활동을 하고 있다. 근대건축이라는 보편적 틀 안에서 멀리 일탈하지도 않으면서도 그 틀을 벗어나려는 강한 비판의 시선과 욕망이 있다. 건축물을 표현하는 수법이 관습적이거나 상투적이지 않고, 현실과의 적절한 관계를 유지하면서 운생동 특유의 차별적이고 도발적인 상상력을 개성 있게 표현하고 있다. 같은 프로그램과 소재를 가지고 출발하더라도 그것을 파악하고 표현하는 방식에 있어 운생동만의 시각으로 표출한다.

그들의 작업 과정은 특정 방식에서 조차 꼭 그래야 한다는, 그렇게 해온 혹은 묵시적으로 동의하는 통례적 건축행위로부터 탈피하면서 관습화된 구태에 새 옷을 입히기 위한 몸부림이다. 기존 체계에서 벗어나 건축의 균질성을 깨뜨리며 시대에 부응하는 질서와 조화라는 맹목적 건축개념보다는 차별성을 통한 작가적 독립성과 개인성이 녹아 있는 또 하나의 지형을 형성하고 있다. 비일상적이며 탈근대적 방식을 통해 관습적 건축과 거리를 두면서, 감성적 표상의 방법을 복합(compound)적이고 메타(meta)적으로 확장해가고 있다.

운생동의 작품에는 개별적 주제의식이 강하게 드러나면서도 공통되게 나타나는 양상이 있다는 점에서 일종의 코드를 이룬다. 표피를 만드는 방식이나 형태공간을 만들어가는 방식이 특히 눈에 띈다. 그들의 표피작업이 갖는 위상은 시각적 관성에 의해 망각되는 일상의 풍경에 낯선 소통의 어휘를 더함으로써 주변과의 상이한 관계 맺기를 하는 것처럼 느껴진다. 주변의 건축 문맥과 차이와 다름을 분명히 하면서 그들만의 정체성을 드러낸다.

도시 내 박스 건물들 사이에 천을 걸친 건물이 들어선 모습을 기억한다. 운생동의 초기 작품인 예화랑이다. 화랑과 전시장, 사무실이라는 기능적 프로그램을 만족시키면서 여러 겹의 천을 늘어뜨려 만든 중층화된 시선의 겹을 느끼게 하는 매력과

independent ways arise, but they also fall, meaning its energy is too weak to form a universal consensus that represents an era or a momentum of change. However, if we were to lock ourselves in the framework of the everyday and ordinary without trying to change within the fence of stereotypes, their voices will not be amplified and communicated throughout society, causing isolation and eventually hindering development.

Even in this situation, there are small yet various voices in Korean modern architecture that practice the change of new architecture. They are creating an exit through the architectural grammar and discourse of the architect as an individual by escaping from the coded idea of everyday life. It is an effort to break away from the regular and repetitive framework that has dominated Korean architecture for a long time. De-ideological and post-modernist vocabulary is used against architecture that has lost its central values.

UnSangDong is at the center of this movement. One can see through UnSangDong's works over the past several years that they are actively working to erase the solid modern framework while pushing forward with everyday work activities. There is strong criticism and desire to escape from the universal framework of modern architecture, without going too far. Their technique of expressing architecture is neither conventional or clich?, but it expresses distinct and provocative ideas with individuality while maintaining an adequate relationship with reality. Even when working with the same program and materials, they understand and express them in their own, unique way.

Their work process is a struggle to put new clothes on the old that have become customs, escaping from common architectural practices that must be, have been or have been implicitly agreed to do so in a certain way. Rather than breaking the homogeneity of architecture by breaking away from the existing system and cultivating a blind architectural concept of order and harmony in response to the times, they are forming another topography where artistial independence and personality are differentiated. Their expansion of the methods of emotional representation is compound and meta, while keeping distance from customary architecture through non-routine and postmodern methods.

The work of UnSangDong has a kind of code in that there is a common aspect that appears while having a strong individual sense of topic. The way they make the skin or form space is especially noticeable. The status of their skin work seems to make a different connection with the surroundings by adding the vocabulary of unfamiliar communication to the daily scenery that is forgotten by visual inertia. They reveal their identity by clarifying the differences within the surrounding architectural contexts.

I remember seeing a cloth-covered building between box-shaped buildings in the city. It was Gallery Yeh, one of the early works of UnSangDong. It satisfies the functional programs of galleries, exhibition halls, and offices, and emits charm that invites the viewer to

발산한다. 몽유도원도 이상봉타워나 크링의 차별된 입면을 보면 관습적 표피의 유형(typology)에 얽매이지 않는 이질적이면서도 복합적인 감성을 담아내고 있다. 또한 성동구 문화센터나 퓨처리즘그리드 미동전자 그리고 Ocean Us에서는 근대건축의 보편적 틀인 그리드 격자 틀을 허물고 있다. 이들 건축작업은 외피에서부터 주변 건물들과는 확연히 대별된다. 구체적인 인식적 그리드는 약화된 모습으로 감추어지고 감각화된 이형이 겉으로 드러난다.

운생동은 여러 공간적 켜가 적층되거나 중첩된 구조를 갖는 공간을 만들기도 하고, 굳어지는 과정에서 공기구멍이 생긴 스위스 치즈(Swiss Cheese)라고 불리는 에멘탈러(Emmentaler)처럼 틀 속에서 빈 공간이 있는 형태를 만들기도 한다. 이러한 형태 조작방식은 파주 유치원에서처럼 공간이 서로 수평 혹은 수직 맞물리거나(interlocked), 성동구 문화센터처럼 내외공간을 상호관입(interpenetrating) 하거나, Skin Scape에서처럼 직선적(rectilinear) 형태로 방향성을 가지며 확장되거나, Press in Paju Book City처럼 나선형으로(spiral) 상승하거나, 그리고 Asian Culture Complex처럼 부지의 평면적 틀을 이형 시키면서 공간을 빼거나 삽입하는 형태로 나타난다. 이 작품들을 보면 운생동이 하고 있는 건축적 얼개를 대략 하게나마 일별하게 해준다.

이러한 형태조작에서 층위의 공간 사이 이동을 중시한 연속체적 공간을 만들어 내기도 한다. 이것을 가능하게 하는 이동동선은 기능적 통로이자 시각적 연속성을 고려한 순환(circulation)의 고리이다. 이 고리는 판에 박힌(routine) 움직임을 생산하기보다는 연속적인 이동이 가능한 흐름을 제공하면서 다양한 시각적 경험을 하게 한다. 이동통로와 그 주변의 공간 배치방식에 따라 사용자에게 공간체험에 대한 자율성을 부여하기도 하지만, 그 반대로 감성이 유린되기도 한다.

베르나르 추미가 언급한 것처럼 움직임을 유도하는 이러한 순차적(sequential) 이동공간은 그 주변에 다양한 이벤트를 만들게 되고, 건물 내·외부와 소통하고 주변환경의 일부가 되기도 한다. 공간의 흐름과 함께 시간의 풍경을 다양하게 지켜봄으로써 일상에서 마주할 수 있는 경험을 극대화하게 된다. 이러한 사고는 최근의 작은 공공건축물에서 잘 구현되고 있다.

공원 초입의 방치된 공간에 계획된 한내 지혜의 숲은 벽에 의한 공간 분할 방식을 사용하여 벽 사이를 이동함에 따라 다양한 공간적 체험이 가능하도록 계획되어 있다. 기존 도서관의 딱딱한 이미지와는 달리 지역주민과 어린이들이 참여하고 함께 만들어가는 "놀이공간"과 같은 도서관을 만들게 된다. 수락행복발전소 또한 스킵 플로어 형식으로 만들어진 중앙의 커뮤니티 공간을 램프로 둘러싸는 형태로, 공간을 움직이면서 지역주민들은 의외의 내부공간을 경험하기도 하며 각자의 관점에 따른 다양한

experience the layers of gazes made by multiple layers of fabric. The distinctive elevation of Mong-yoo-do-won-do Lee Sang-bong Tower and Kring Kumho Culture Complex shows the heterogeneous and complex sensibility that are not tied to the typology of a conventional outer layer. Also, Seongdong-gu Cultural Center, Futurism Grid, Midong Electronics HQ and Ocean Us also break down the grid framework that is the universal framework of modern architecture. Even the outer skin of these architectural works are distinctly set apart from the surrounding buildings. The concrete cognitive grid is concealed in its weakened form, and the sensibilized, heteromorphic form is revealed.

UnSangDong makes spaces in which several spatial layers are stacked or overlapped, as well as forms like Emmentaler, a Swiss Cheese that gains air holes in the hardening process, with an empty space in the frame. The space in this type of form fabrication method can be applied in the same way as in Paju kindergarten, where the spaces are horizontally or vertically interlocked; interpenetrate the inner and outer spaces as in Seongdong-gu Cultural Center; expand with rectilinear shape as in Skin Scape; go up in a spiral as in Press in Paju Book City; or by subtracting or inserting space by releasing a flat frame of the site as in the Asian Culture Complex. A study of these projects reveals a quick glance into the architectural framework that UnSangDong pursues.

This type of manipulation also creates a continuous space with an emphasis on movement between layers of space. The circulation plan that makes this possible is a circulation loop that takes into account functional pathways and visual continuity. Rather than producing routine movement, this loop allows a variety of visual experiences while providing a flow that allows continuous movement. Depending on the circulation path and the arrangement of space around it, users are either given autonomy about the experience of space or violated of their sensibility.

As Bernard Chumi mentioned, this sequential circulation space that drives movement creates various events around it, communicates with the building's interior and exterior, and becomes part of the surrounding environment. By watching various landscapes of time with the flow of space, we maximize the experiences we can face in everyday life. Such thought processes are well implemented in recent small public buildings.

The Hannae Forest of Wisdom was planned for the uninhabited space at the beginning of the park. Using a space partition method with walls, the area was designed to enable various spatial experiences as visitors move between the barriers. Moving away from the existing notion of solemn libraries, they created a library that would be a "play space" where local residents and children could participate and create together. Surak Happiness Powerhouse also has a central community space built in the form of a skip floor that is surrounded with lamps. As they explore the space, local residents happen upon unexpected internal spaces

방식의 시각적, 공간적 체험을 하게 된다. 제한된 공간 내에서 관찰과 체험을 통한 다양한 지각적 경험과 주관적 인식이 지역주민들로 하여금 건축공간에 대한 풍성한 감정이입(empathy)을 자연스럽게 이끌어 낸다.

    흥미롭게도 이러한 운생동의 표현은 다른 건축작품에서 규칙적이고 반복적으로 나타나지 않는다. 이는 각 작품에 사용하는 그들의 어휘가 자기복제적 답습이 아니라, 새로운 탐험이기 때문이다. 빈약한 창작작업이 생존을 위해서 무리하게 숨을 내쉬는 억척스러움 보다는 창작의 즐거움을 느끼는 여유가 있어서이지 않을까 싶다. 작업 자체가 어떤 놀이에 가까운 그들의 작업은 형식적 공간을 만들어가는 논리보다는 공간형태적 영감에 더 힘이 실린 방식이다. 다른 여러 건축물에서 마찬가지로 명쾌하고 깔끔하기 보다는 분주하고 복잡한 형태는 기하학적 용어로 쉽게 설명하기 어려운 자기표현적 몸짓이다.

    현대건축은 과거와는 달리 기능적 배열에만 충실하거나 고전적 어휘나 양식으로 도배되지 않는다. 위계나 체계보다는 덜 구속된 탈형식적이며 그 어휘들은 다른 분야에서 차용되거나, 건축가 개인의 주관적 어휘들로 구성되어 보편화 되어가고 있다. 그런 측면에서 운생동에서 쓰는 건축적 어휘들 또한 새롭다기 보다는 현대 건축에 차용된 익숙한 용어들이다. 그럼에도 불구하고 디자인을 만들어 가는 독특한 발상이 눈에 띄며, 여기저기서 복제한 디자인들과 같은 콜라주 기법보다는 훨씬 건강하다. 어찌 보면 거칠게 보일 수 있고 현실과의 유리된 전위적 디자인처럼 보일 수도 있으나, 창작적 영감과 사고의 요지를 표출하는 특유의 방식으로도 설명될 수 있다. 매해 넘치도록 생산되는 건축물들을 보면서, 이젠 혼이 깃들거나 장인정신이 베어있는 건축물들이 서울의 하늘아래에 등장할 때가 아닌가 곱씹어 보기도 한다.

    운생동이 들고나온 복합체라는 어휘의 전개방향은 다양한 어휘를 실험해보는 과정의 결과물로 읽힌다. 실험은 아직 진행 중이며 디자인은 진화하고 있다. 이것이 지금의 운생동보다 향후가 기대되는 이유이다. 관습적으로 굳어진 근대적 표현이 아니라 운생동만의 표현의 혁신을 이끌어, 누가 보더라도 운생동스러운 (UnSangDong-resque) 건축을 만들어가기를 기대한다.

and have various visual and spatial experiences according to their own perspectives. Different perceptual experiences and subjective perceptions through observing and experiencing a limited space naturally brings out deeper empathy of architectural spaces in the local residents.

    Interestingly, this expression of UnSangDong does not appear regularly or repetitively in their other architectural works. This is because their vocabularies for each project are not self-replicating, but new explorations. I think this is possible because they can afford to feel the joy of creating rather than letting creative work become unnecessarily suppressed under the burden of survival. Their work, which is closer to a type of play, chooses to empower space-form inspiration rather than the logic which creates a formal space. As in many other buildings, busy and complex forms rather than clear and neat forms are self-expressive gestures that are difficult to explain easily in geometric terms.

    Contemporary architecture is not only devoted to functional arrangement like in the past, nor is it dominated by classical vocabulary or style. It is less formalized than hierarchical or systematic, and its vocabularies are borrowed from other fields, or composed of the subjective vocabulary of individual architects and then universalized. In that sense, architectural vocabularies used by UnSangDong are not new but familiar terms borrowed from modern architecture. Nonetheless, their unique process of creating designs stands out, and is a much healthier methodology than collage techniques that replicate designs from here and there. It may seem rough in some ways and appear to be avant-garde design that has lost touch with reality, but it can be described as a unique expression of the essence of creative inspiration and thought. Looking at the number of overproduced buildings every year, I wonder if it is not now time for architecture that bears soul and craftsmanship to start appearing under the skies of Seoul.

    The development direction of the word "complex compound" by UnSangDong is read as a result of the process of experimenting with various vocabularies. The experiment is still ongoing and the design is evolving. This is why we look forward to the future of UnSangDong, more than its present. I look forward to seeing the innovation of UnSangDong's unique expression to create architecture that is not just relying on customary modern expression but rather that which is UnSangDong-resque in every way.

2012 Yeosu EXPO
Hyundai Motor Group Pavillion

2012 여수 엑스포 현대자동차 그룹관

Location: Site in Expo 2012 Yeosu, Jeonnam, , Republic of Korea **Site Area:** 1,960 m² **Building Area:** 1,397.50 m² **Total Floor Area:** 2,334.81 m² **Designer:** Kim Sung Min, Kim Min Tae, Hyun Sang Heon, Goh Young Dong, Kang Seung Hyun, Jang Chol Min, Kim Mi Jung, Son Min Sun, Kim Hye Soo **Client:** Innocean Worldwide **Structure:** Steel Frame Construction **Photographer:** Jaekyeong Kim(Model), Sergio Pirrone

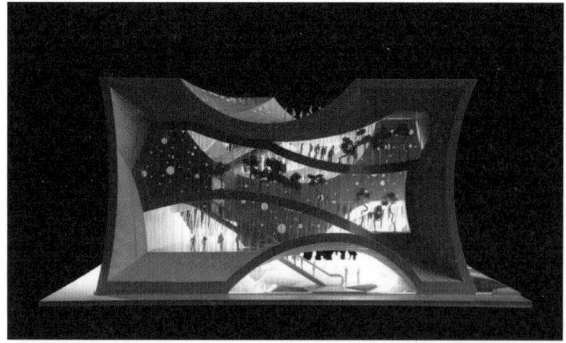
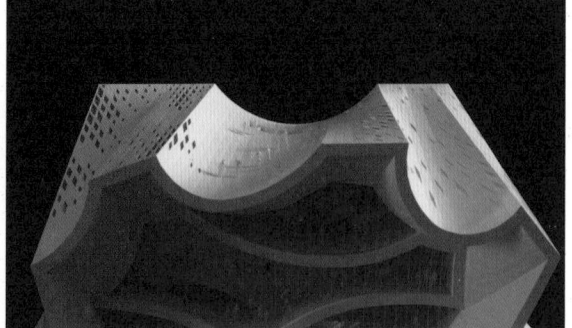

여수엑스포 현대자동차그룹관을 통하여 인간역사의 이동수단의 혁신을 만들어내기 위한 기업의 브랜드이미지를 건축적으로 변환시키는 작업을 제안하려고 하였다. Motion Imagination라는 개념을 통해서 현대자동차그룹의 움직임(Motion)과 미래의 새로운 물결의 상상력(Imagination)을 결합한 건축을 재현한다. 이는 지금까지 성취되지 않았던 미래적 상상을 더욱 증폭시키는 가능성을 재현한데 있다. 자연과 도시에 대한 무한한 가능성과 상상력을 현실로 변화시키는 힘을 재현하여 현대 자동차그룹의 정신과 첨단기술의 가능성을 표현한다. '이동성 motion'이라는 건축적 주제를 역동적 단면의 새로운 구성을 통하여 제안하려고 한다. 이는 Blue Ocean Wave를 의미하는 입체적 파동이 만들어내는 다양한 움직임의 단면을 겹치고 연결시켜 고정되지 않고 항상 변화하는 현대자동차그룹의 브랜드이미지를 건축적으로 표현하는 것과도 연결된다.

Though Yeosu EXPO Hyundai Motor Group Pavilion, Architectural work is suggested to convert the corporate brand image, which is creating the innovation of transportation in human history. This architecture is combined with imagination of new future waves and the motion of Hyundai Motors Group. It is also reproduced though the concept of 'Motion Imagination'. This is reproducing possibility of amplifying the futuristic imagination which is never achieved until now. The possibility of high-tech and spirit of Hyundai Motor Group is expressed by reproducing the power. It actualizes the imagination and infinite possibilities of nature and urban. The architectural issue of 'motion' is proposed through the new formation of dynamic section. The section of various waves, which signifies three dimensional "Blue Ocean Wave" movements which is overlapped and connected. This also connects with the architectural expression of constantly changing corporate brand image.

House ONE : Chronotope Wall House
크로노토프 월 하우스

Location: Pangyo-dong 571, Bundang-gu, Seongnam-si, Gyeonggi-do, Republic of Korea **Completion:** 03, 2016
Use: House **Site Area:** 255.2m² **Building Area:** 125.65m² **Building Coverage ratio:** 49.23% **Gross Area:** 228.78m²
Building Scale: B1 ~ 2F **Structure:** Reinforced Concrete **Height:** 11m **Design team:** Kim Mi Jung, Kim Min Kyun
Photographer: Jaekyeong Kim(Model), Sergio Pirrone

시간은 공간에 갇혀 있고, 공간은 시간을 잃어버렸다. 그러나 시간과 공간은 혼자 존재할 수 없다. 인간도 혼자 존재할 수 없다. 가장 구조적이며 건축적인 어휘인 월(Wall)을 통하여 주거를 구성하려 했다. 월은 공간과 공간을 가로지르며 단절되고 잘려진 장소를 구축하는 특성을 가지지만, 크로노토프적 설정에 의해서 변화된 경험과 상황을 생성해낼 수 있다. 크로노토프는 여러 지표간의 융합과 축의 교차를 통하여 공간과 시간을 통합하는 하나의 방식이라 볼 수 있다. 결국 건축적으로 차용된 월을 단절과 분할의 구조로 사용하는 것이 아니라 반대적인 태도인 통합과 연속, 동시적 공간과 연속적 시간의 틀로 사용하려 했다. 이는 현대사회의 가족의 단절과 해체 상황을 월의 단절로 인식하고, 월의 변형을 통하여 소통하는 가족으로 다시 되돌려 놓는 인간관계의 회복적인 공간구조를 재현하려 했다.

Time is locked in space, and space lost time. However, neither time and space can exist without the other. So does human being as well. We are going to build a living space using the most structural and architectural terminology, 'wall'. Even though a wall traverses between spaces and constructs an isolated area, it can generate an experience and a situation that are altered by the Chronotope. Chronotope can be seen as a method to unite space and time through the fusion of various indices and the intersection of axes. We are eventually about to use a wall, which is being used architecturally, not as a structural element for severance and division but as an element for integration, continuity, coincidental space, and continuous time. This is that the severance of a family in modern society is recognized as the severance of a wall. We are going to reproduce a restorative spatial structure of human relationships that restore the severance of a family into an interactive family by altering a wall.

An Architect that Can Change
변화 가능한 건축가

Seungman Baek (School of Architecture, Yeungnam University)
백승만 (영남대학교 건축학부)

(사)한국건축설계학회는 건축갤러리의 2018 초대건축가 작품전으로 더시스템랩(05.25-06.22)에 이어 운생동 건축전(10.19-11.16)을 개최한다. 그동안 한국에서 특별한 회고전을 제외하고는 동시대의 건축가에 대한 개인전은 매우 드물었다. 유사하지만 여러 모로 양상을 달리하는 한국현대미술계와의 차이가 여기에서도 존재한다. 이러한 맥락에서 앞으로 한국현대건축의 비전을 제시할 건축전문갤러리가 건축설계학회에 의해 발족된 것은 그 의미가 크다.

운생동이 널리 알려지게 된 계기를 마련한 작품은 많은 건축전문지에 소개되었던 예화랑(2006)이다. 프랑스 건축전문지 d'Architectures 편집장을 엮임하고 프랑스건축연구소 IFA의 책임자였던 프랑시스 랑베르(Francis Rambert, 1954-)가 한국을 방문했을 때, 유난히 예화랑을 보고자 했던 것이 기억난다. 그 후, 크링(2008)과 더힐갤러리(2011)를 몇 차례 방문할 기회가 있어서 운생동 건축을 잘 살펴볼 수 있었다. 매우 과감하면서도 세심한 디테일, 독특한 현대적 감각 등이 인상적이다. 오늘날 운생동 건축이 '특이하다'라고 평가되고, 혹자는 파격적인 건물을 일컬어 '운생동스럽다'라고 표현하기도 한다. 그 의미가 긍정적이던 부정적이던 운생동 건축은 그들만의 특성을 인정받고 있다.

Following an exhibition by The_System Lab (25 May-22 June), the Architecture Design Institute of Korea (ADIK) will hold the UnSangDong Architecture Exhibition (19 Oct-16 Nov) at the Architecture Gallery as the 2018 Invited Architects Exhibition. Until now, solo exhibitions of contemporary Korean architects, besides the occasional retrospective exhibitions, have been very rare. This is a notable difference from the Korean contemporary art world, which seems similar to contemporary architecture but actually differs in many ways. In this context, it is meaningful that the Architecture Gallery, which will present the future vision of modern Korean architecture, was set up by ADIK.

The project that put UnSangDong on the map is Yeh Gallery (2006), which was introduced in many architectural journals. I remember that when Francis Rambert (1954-), who was the editor of the French architectural magazine d'Architectures and head of the French Institute of Architecture (IFA), visited Korea, he was especially keen on visiting the Yeh Gallery. After that, I had a chance to visit Kring Kumho Culture Complex (2008) and The Hill Gallery (2011) several times, so I was able to take a good look at UnSangDong architecture. It is very bold but has carefully thought-out detail and a unique modern sense. Today, the architecture of USDSpace is evaluated as 'unusual',

한국의 현대건축에 대해서는 김수근(1931-86)과 김중업(1922-88)의 양대 줄기를 언급하지 않을 수 없다. 각각은 해외에서의 경험이 그들의 건축에 많은 영감을 주었고 프로젝트를 통하여 한국성을 찾으려고 끊임없이 노력을 하였다. 이러한 노력은 그 다음 세대에 이어졌으며, 그동안 한국 현대건축의 결실과 기성세대의 주류를 이루었다. 그러나 90년대 IMF 외환위기를 겪으면서 건축분야 또한 어려워지고 사회에서 독립적인 건축가들이 사라질 즈음, 새로운 세대가 나타나기 시작하였다. 굳이 한국적 해석을 언급하려하지 않거나 '가장 한국적인 것이 가장 세계적인 것이 될 수 있다'는 자신감을 드러내는 건축의 X세대이다. 이는 IMF 외환위기로 인해 한국 보다는 세계를 대상으로 치열하게 생존한 신세대 건축가의 선택으로 보인다. 보다 창의적이고 진취적인 X세대[1] ADIK(Architecture Design Institute of Korea) 건축갤러리에서는 이러한 신세대의 건축가를 주목하며, 운생동이 그 대표적인 건축가에 해당된다.

운생동과의 직접적인 만남은 '2014 바이오디지털 예술·건축전'을 계기로 이루어졌다. 2011년 파리에서 프랑스 교수들로부터 Bio-Digital City 국제워크숍을 제안 받았을 때, 'Bio-Digital'이란 용어자체에 매우 흥미를 느꼈다. 생물학적인 것/자연(Bio)과 시대의 기술(Digital)이 융합된 결과물은 동시대적이며 장소적 제약 없이 평가되고 발전될 수 있다. 특히 자연에 대해서는 인간의 본질을 내포하고 있으므로, 모든 예술의 영원한 주제가 될 수 있으며, 그것이 이성적이던 감성적이던 시대의 기술과 부합하여 표현된다. 2014년과 2017년의 전시회에 함께 하였던 더시스템랩(The_System Lab), 스케일(SCALe)과 운생동은 서로 궤를 달리한다. 더시스템랩과 스케일은 시대의 기술(컴퓨터)을 기반으로 자연성을 표현한 반면에, 운생동은 자연의 철학을 바탕으로 컴퓨터를 포함한 시대의 기술적 표현을 하고 있다. 사무실 명칭 또한 X세대의 특징이라 할 수 있는 국제화를 표명하는 영문조합어가 아닌, 고전적인 동양철학의 기운생동(氣韻生動)에서 유래한다. 이러한 뚜렷한 주관과 동시대성에서 가장 동양적인 것이 세계적으로 가장 주목받을 수 있겠다는 생각이 들었다.

운생동은 장윤규건축실험아틀리에(1997)를 거쳐 신창훈과 함께 설립된 즈음(2003) 이미 건축철학이 확고히 잡힌 듯하다. 영어로도, 한글로도 그 의미가 전달되기 힘든 '운생동'(韻生動)은 지극히 고전적이다. 국제화시대에 어울리지 않는 언어의 선택은 오히려 건축가의 철학을 강조하고 있다. '생동하는 힘'(氣), 즉 에너지에 대한 표현이 건축구성의 궁극적인 목표로 보인다. 건축가의 간결한 먹물 스케치에서는 동양적 기운이 느껴진다. 운생동 건축의 주요 개념으로 언급되는 '복합체'(Compound Body)는 기운생동의 화법(천지 만물이 지니는 생생한 느낌을 표현한다)을 현대적 건축용어로 번안된

and some refer to unconventional buildings as 'UnSangDong-resque'. Whether this saying has a positive or negative connotation, the architecture of UnSangDong is being recognized for its uniqueness.

One cannot discuss Korean modern architecture, without mentioning the two master architects, Kim Swoo-geun (1931-86) and Kim Joong-up (1922-88). Each was inspired by his experience abroad, and they constantly strived to recreate Koreanness through their projects. This effort carried on to the next generation, and which has bore fruit as modern Korean architecture and became mainstream in the older generation. However, around the time when the IMF financial crisis of the 1990's had a negative impact on the construction sector and independent architects started to diminish from society, a new generation emerged. They were the Generation X of architecture, either distancing themselves from Korean interpretations or confident that 'the most Korean ideas can be the most international ones.' This seems to be a deliberate direction taken by the new generation of architects who survived the financial crisis by outperforming in the global market; the extra creative and adventurous generation X.[1] The Architecture Gallery at ADIK focuses on this new generation of architects, and UnSangDong is at the forefront of this group.

I had the opportunity to personally connect with UnSangDong through the '2014 Bio Digital Art and Architecture Exhibition'. When French professors proposed a Bio-Digital City International Workshop in Paris in 2011, I was quite intrigued by the term 'Bio-Digital' itself. The converged result of 'biological/natural (bio)' and 'technology of the age (digital)' is a contemporary one and can be evaluated and developed without restriction of place. Nature, as it contains the essence of man, can be an eternal subject of all arts and is expressed rationally or emotionally in accordance with contemporary technology. UnSangDong chooses a different direction than the other two firms, The_System Lab and SCALe, which each held exhibitions in 2014 and 2017. While The_System Lab and SCALe express naturalness based on the technology of the times (the computer), UnSangDong uses the technological expression of the era, including computers, based on the philosophy of nature. The name of the office is also derived from classical oriental philosophy, gi-un-saeng-dong (qìyùn shēngdòng; vivid capture of rhythm or spirit), rather than an English word combination expressing internationalization, which is characteristic of the generation X. It came to me that such distinctly oriental values with clear subjectivity and contemporaneity could attract the most attention worldwide.

UnSangDong seems to have firmly established its architectural philosophy since its beginning as Jang Yoon-Gyoo Architectural Experiment Atelier (1997) and foundation with Shin Chang-hoon (2003). 'Un-Sang-Dong', a difficult concept to convey in English or Korean, is a very classic term. Choosing language that goes against the era of internationalization further emphasizes the philosophy of

듯하다. 많은 비평가들이 운생동 건축의 특이성(singularity)을 언급하게 되는 요인은 건축가의 철학적 개념이 프로젝트마다 다른 현상으로 재현되기 때문일 것이다. 만일 운생동이 하나의 건축구성법(기술)을 연마하였다면, 건물은 점차 세련되는 반면에 그 특이성은 감소될 수 있다.

운생동은 동시대의 기술에 대하여 매우 유연한 자세를 지니고 있다. 장윤규는 2004년 5월호 공간지에서 '인터랙티브 맵'(Interactive map)을 건축개념으로 제시하면서 '건축가의 사전'이란 제목으로 운생동의 구체적인 건축구성방법론(Metamorphose, one+one, Floating+floating, Transprogramming, Reaction body, Compound body, Skinscape, Voidscape)을 설명하였다. 들뢰즈(Gilles Deleuze), 바르트(Roland Barth), 코라(chora) 등 현학적인 서양철학을 '기' 동양철학과 가로지르기를 하면서 현대적 건축언어를 운생동화 하고 있다. 그들의 건축언어는 자연의 이치에 속하는 우주론적(존재론적) 철학을 바탕으로 때로는 서양철학에서, 때로는 현대건축방법론에서, 때로는 예술가와의 협업에서 재구성되기에, 항상 실험적이며 지극히 현대적인 표현이 가능해보인다. 최근에 발간된 이태리 건축전문지 l'Arca International의 'Compound Body; UnSangDong Architects'(2017)에서는 4가지 유형(Becoming Animal, Skin Scape, Clip City, Mythological Imagination)으로 운생동 건축을 재분류하여 설명하고 있다. 위의 4가지 유형은 르 꼬르뷔제가 제시한 근대건축의 4가지 구성(비교적 쉬운, 매우 어려운, 매우 쉬운, 매우 일반적인 유형)과는 달리 유동적이다. 만일 운생동 건축이 규범화된다면, 그 생명력을 잃을 수도 있다. 성수문화복지회관(2013), 몽유도원도 이상봉타워(2018) 등은 규범화되지 않았기에 실현될 수 있다. 건축설계에 있어서 철학을 먼저 세우고 방법론을 찾는 것은 매우 힘든 일이며, 드문 경우이다. 그러나 운생동 건축은 이미 실험기를 지나 전성기를 맞이하고 있으며, 앞으로도 동시대의 문화적, 기술적 통섭이 끊임없이 진행된다면, 그들만의 특이성은 더욱 빛날 것이다.

the architect. The expression of 'living force (i.e., energy) seems to be the ultimate goal of their architectural composition. The architect's simple ink sketches hold an oriental vitality. The 'Compound Body', which is referred to as the main concept of UnSangDong architecture, seems to be a reinterpretation of an existing concept (expressing the vivid feeling of all things) into a modern architectural term. Many critics point to the singularity of UnSangDong architecture because the architect's philosophy is recreated as a different phenomenon in each project. If UnSangDong had cultivated a singular building construction technique, the building may have been refined, but its specificity would have been reduced.

UnSangDong has a very flexible attitude toward contemporary technology. Jang Yoon-Gyoo proposed 'interactive map' as an architectural concept in the Space magazine's May, 2004 edition and shared the architectural composition methodology (metamorphose, one+one, floating+floating, transprogramming, reaction body, compound body, skinscape, voidscape) of UnSangDong under the title, 'Architect's Dictionary'. They assimilate modern architectural language into their own UnSangDong style by intertwining pedantic Western philosophy such as Gilles Deleuze, Roland Barth, and chora with the 'ancient' oriental philosophy. Their architectural language is based on a cosmological philosophy belonging to the laws of nature. And as the language is sometimes reconfigured in Western philosophy or modern architectural methodology, and other times in collaboration with artists, their expressions always manage to stay experimental and modern. The recently published Italian architectural magazine l'Arca International's' Compound Body, UnSangDong Architects (2017) reclassifies and restructures UnSangDong architecture into four types (Becoming Animal, Skin Scape, Clip City, and Mythological Imagination). These four types are fluid, unlike "The 4 Compositions" of modern architecture proposed by Le Corbusier (rather easy, very difficult, very easy, very general). If UnSangDong architecture is normalized, it may lose its vitality. Seongsu Cultural Welfare Center (2013), Mong-yoo-do-won-do Lee Sang-Bong Tower (2018) were realized because they were not standardized. Figuring out the methodology after first deciding on a design philosophy is very difficult and rare in architectural design. However, the architecture of UnSangDong is already coming to its prime after passing through an experimental period, and with a continuous consilience of culture and technology, its singularity will shine even brighter.

1 캐나다 작가 더글러스 쿠플랜드의 소설 <Gerneration X: Tales for an Accelerated Culture>에서 유래된 단어로서, X는 '정의할 수 없음'을 의미하고, X세대는 이전 세대의 가치관과 문화를 거부하는 이질적 집단이다.

1 The word was originally used in the novel 'Generation X: Tales for an Accelerated Culture' by Canadian author, Douglas Coupland. X represents 'that which cannot be defined', and Generation X is a disparate group which refuses to accept the values and culture of their former generation.

Gallery Yeh
예화랑

Location: nam-gu, Seoul, Republic of Korea **Use:** Mixed-use facility **Site area:** 567.5m² **Building** **Gross floor area:** 1,995.14m² **Scale:** B2 ~ 7F **Construction:** Gujin Industrial Development **Client:** Lee Sook Young **Exterior finishing** T50 base panel, Exposed concrete, T24 transparent **Finishing:** Epoxy coating, Base panel, Exposed concrete **Photographer:** Jaekyeong Kim

예화랑에 적용된 스킨스케이프의 개념은 '스킨이 공간되기'라 볼 수 있다. 스킨스케이프의 막은 전시의 정보를 제공하는 미디어가 되어 외부전시를 이끄는 거대한 캔버스가 된다. 스킨스케이프의 공간을 여러겹의 스킨을 관통하는 특이한 공간 경험을 제공한다. 사람들은 이 공간 사이를 경험하며 벽 속에 숨어 있는 새로운 공간의 질을 발견한다. '갑옷의 틈새'와 같이 스킨은 단순히 공간을 한정짓기 위한 표피가 아니라 공기적 틈새를 가지며 내외부의 모호한 경계를 이끄는 가벼운 물성의 표피이다. 무거운 언어를 가벼운 재료와 구조의 새로운 공간화에 의해서 형성하여 그 속에 은유적인 아이러니와 메타포가 숨어 있다. 가벼움을 통해서 획득되는 시학적 무거움의 틀을 구성한다. 스킨의 질은 틈새의 가능성과 연관된다. 틈새의 스케이프를 만드는 방식을 제안하고 그 안에 숨겨진 공간적 풍부함을 도출한다.

The concept applied in 'Gallery Yeh' can be categorized as 'Spatialization of Skin' and/or 'Mediazation of Ski' - screen for the skinscape can be the medium to provide exhibit information as well as the huge canvas attracting outside events. Space for the skinscape offers unique spatial experience of puncturing through multiple layers of skin, in which each of its layers come as different spatial quality. 'Spatial Surfing', 'Skin Surfing', 'Pictorial Surfing', 'Organizational Surfing' are some of the codes appearing along such experience, while each surfing twists and intertwines to create spatial complexity as a whole.
 Like 'Crack of Armor', skin is not the surface that envelopes the space, but is bound by air that is light material and metaphorical interpretation and irony behind the new possibilities of space. The framework of heaviness is gained through the lightness, the quality of the skin is linked with the possibilities of creating gaps, so thus the space become enriching experience of discovering the hidden layers of logic and irony. thus the space become enriching experience of discovering the hidden layers of logic and irony.

50
51

Gallery The Hill
갤러리 더 힐

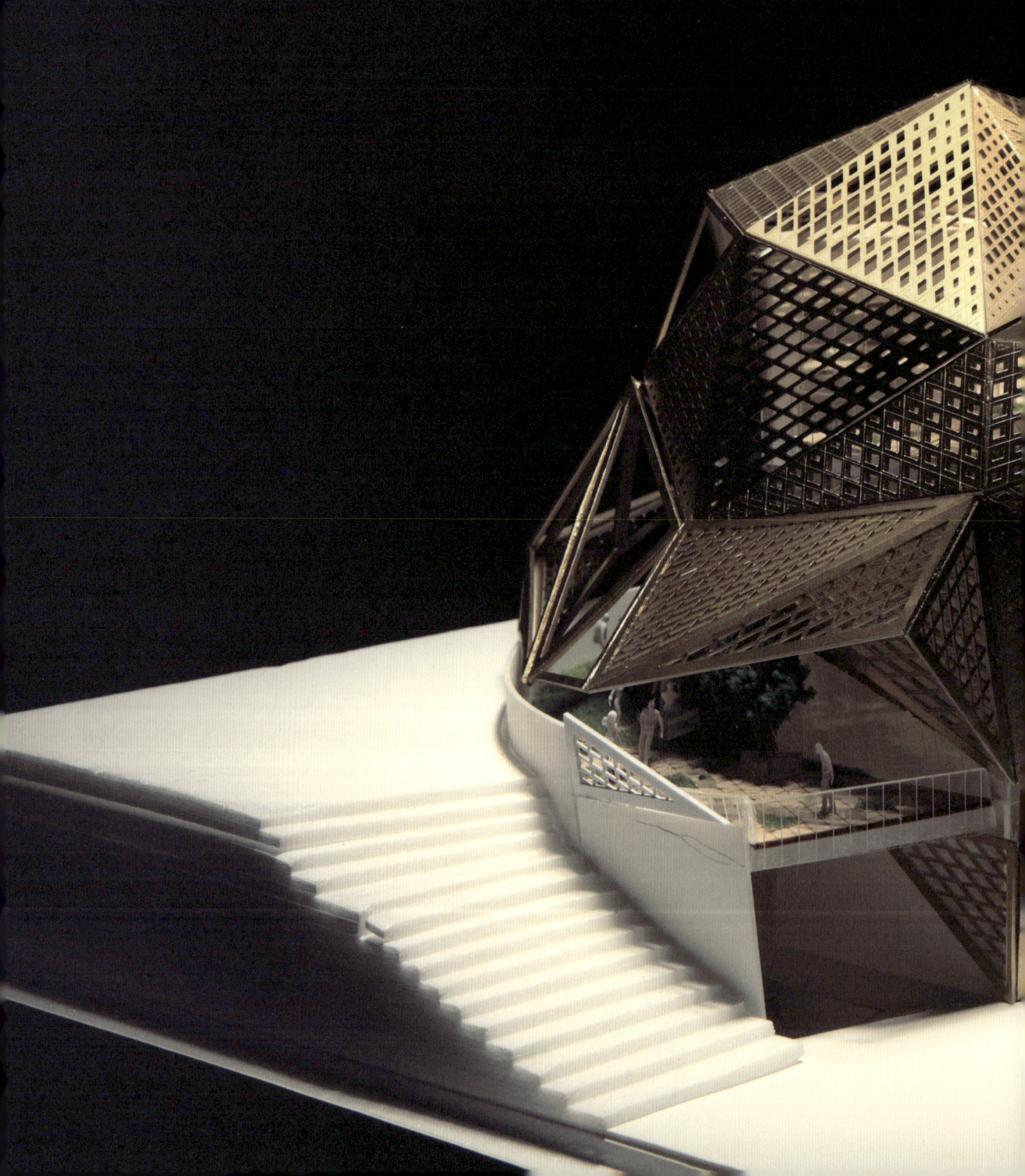

Location: 810, Hannam-dong, Yongsan-gu, Seoul, Republic of Korea **Building Area:** 485.076 m² **Total Floor Area:** 2,730.31 m² **Architects:** Jang Yoon Gyoo, Shin Chang Hoon, Kim Kyeong Tae **Designer:** Sim Jehyun, Choi Young Eun, Ahn Bo Young, Yang Ki Ran **Client:** Kim Sang Woon_Hansjaram Corporation **Structure:** RC Structure **Photographer:** Jaekyeong Kim(Model), Sergio Pirrone

한남동 주거단지의 커뮤니티시설중의 하나인 근린생활시설건축을 대지적 지형으로 치환하는 작업으로 더힐갤러리를 제안한다. 서울에서 아파트 근린생활시설은 주거시설에 비해서 디자인에 대한 비중을 주지 않는 주변의 건축으로 치부되고 있다. 우리는 이번 프로젝트를 통하여 이 주변적 건축을 새로운 커뮤니티 프로그램적 적절성의 역할을 복원하는 시도로 삼고자 한다. 대규모 주거단지를 구성하면서 기존의 지형의 파괴로 만들어진 자연적 단절을 새로운 지형의 맵을 이용하여 환원하는 과정을 가진다. 자연을 연결하는 인공산과 같은 도시적 역할을 구성한다. 기존의 더힐 아파트단지와는 확연하게 구분되는 별종의 아웃라이어(outlier)와 같은 존재이며 대지의 형상을 삼각형의 조합으로 이루어진 추상적인 스킨으로 변환시킨다. 크리스토의 작업과 같이 기존의 건물을 새로운 막으로 덮어내고 새롭게 탄생시키는 일련의 대지예술과 같은 방식을 개입하였다. 쭈굴쭈굴한 힌색막을 기하학적인 건축어휘로 변환시켜 크리스탈과 같은 공간을 내외부에 완성한다. 랜드스케이프이며 빛의 산으로서의 랜드마크가 된다. 경사를 따라서 올라가는 대지의 프레임은 축제와 이벤트를 수용하는 다양한 프로그램의 장치이며 내외부의 이중막으로 작용하여 풍경과 환경을 제어하는 중간 매개체의 공간을 구성한다.

We suggest The Hill Gallery, a community facility in the housing district at Hannam-dong, by replacing the neighborhood facility to the topography. The neighborhood facility is considered a subordinate architecture which doesn't care much about its design. Our aim is to restore the role of this subordinate building as the local community program. In the process of composing a large scale housing block, severance of nature which happened as a result of topographic destruction is reverted using a new topographic map. It acts like an artificial mountain connecting nature and the urban. The shape of the site is converted to an abstract skin consisted of a triangular pattern. It becomes the landscape and a land mark as a light mountain. The frame of the site rising along the slope accommodates festivals and events. A half transparent skin is suggested which changes according to the change of the inner programs and the functions. Leaning skin and the outdoor frame consist of the dynamic lighting mountain. The landscape, locating partial void including an eco deck, is a green space in the artificial mountain. It also plays a role of a gate and a link between the city and housing. As an urban connection approaching the city, it provides cultural space for example, a cultural vision deck, recording the memory of Hannam housing and being the observatory for the community.

Hayub Song (Department of Architecture, Chung-Ang University)
송하엽 (중앙대학교 건축학과)

운생동건축사사무소(이하 운생동)의 건축은 운생동 초기에 디자인했던 갤러리와 도시 건물의 모습과 크게 다르지 않다. 초기에 건축저널을 통해 발표했던 장윤규의 실험적 텍스트는 1990년대 해체주의에서 시작하여 건축 디자인과 관계되는 방법론을 찾기 위한 부단한 노력이었다. 1990년대 이후 유학을 마치고 돌아온 건축가들이 쏟아내는 텍스트들이 크게 다르지 않은, 토종 건축가 장윤규의 이론 섭렵은 유학을 가지 않고도 해외 수준에 버금가는 건축적 사고를 진행할 수 있는 가능성을 보여주기에 충분했다. 그의 열정은 글과 작품이 실린 건축저널(설계 작품과 공모안들로 채워진 지금의 저널들과 달리, 2000년대 초반의 건축저널은 아직 학생들이 공부하기에 충분한 텍스트들이 많이 실렸다)을 보는 학생들과 건축설계에 몰입한 젊은 건축사보들에게 많은 희망을 준 것도 사실이다.

        운생동의 작품들은 그들이 공동대표로서 사무실을 시작한 2000년 이후 특이성이 돋보였다. '특이성'이란 표현은 천의영(경기대학교 건축전문대학원 교수)이 작성한 비평문에서 운생동 작업에 명명한 특성이다(『SPACE(공간)』 560호 참고). 특이(Singular)하다는 표현은 말 그대로 개별자적이며 독창적이라는 뜻으로 나름 운생동 특유의 스타일을 이뤘다는 것이다.

The architecture of UNSANGDONG architects is not markedly different from the gallery and buildings of its early designs. The experimental texts of Jang Yoongyoo, which were presented in the architectural journals of the early days, were an expression of his perseverance to discover a methodology related to architectural design, and those articles were not in conflict with the articles written by architects who came back from their studies abroad in the 1990s. Jang Yoongyoo's familiarity with various critical and cultural theories proved that it was possible to engage in architectural theoretical thinking on an international level without having studied abroad. It is also a fact that his passion and his journal articles (unlike those found in contemporary journals filled with design projects and competition proposals, the architectural journals of the early 2000s published many informed articles for students) inspired many students and budding architecture companies with hope.

        The architecture of UNSANGDONG, since opening in 2000 as co-representative, exhibited a sense of 'singularity'. The attribute 'singularity' is a property coined by Chun Euiyoung (professor, Kyonggi University) to describe the works of UNSANGDONG (covered in SPACE, issue 560). As the word 'singular' denotes a certain distinctiveness and creativity, it suggests that UNSANGDONG developed its own unique style.

'예화랑'(2005)은 운생동의 작업을 선두적 반열에 올려놓은 대표적 작품이다. '도시 캔버스'라고 불리는, 자유롭게 각진 형태의 콘크리트 외벽들이 서로 틈을 가지며 복잡다단한 주변 건물의 배경이 되고 있다. 특이성은 1998년에 미니멀하게 등장했던 김인철의 '김옥길 기념관'과 같이, 각진 콘크리트 벽이 주는 2000년대의 도시 건축물로서의 순수추상적인(pure abstract) 모습을 띄고 있다. 기존의 노출콘크리트 벽이 수직으로 선 모습이었다면, 예화랑은 자유로운 각으로 되어 무심한 듯 자연스런 비례를 보이고 있다. 불규칙한 기하학으로 파격을 주는 기법은 장윤규와 신창훈이 만난 건축사사무소 아르텍의 김관석에게 조형적으로 영향을 받은 것이라 해도 과언이 아니다. 1990년대 김관석의 건축은 입면과 조형에서 자의적인 각진 기하학을 표면이나 매스로 강조하며 나름 파격을 주었다. 그러나 소위 집장사 건물들도 쉽게 따라할 수 있는 '간편함'이었다. 입면의 자유로움을 추구하는 인식의 변화는 이끌었지만, 건축가의 깊은 성찰을 보여주지 못했다. 이와는 달리 예화랑은 구조적인 변화를 입면의 자유로움으로 이끌어내어 쉽게 따라하기 어려운 도시 캔버스로서의 역할을 하고 있다. 구조적 잉여가 표면의 자유를 이끌어낸 형국이다.

예화랑이 도시에 생동감을 더하는 예술섬으로서 주변의 도시 상황과는 다름을 강조했다면, '성동문화복지센터'(2013)는 열린 공공건물로서의 이미지를 주기에 충분했다. 내부계단의 움직임과 구조의 사선이 교묘히 생동감 있게 열린 입체적인 도시 이미지를 만들었다. 운생동이 이론적으로 주창했던 '복합체'라는 뜻이 잘 어울리게 건물의 실내와 실외를 통하여 사람들의 움직임이 관찰되는 형국을 이끌어내었다. 도시 현상을 면밀히 관찰하여 얻어낸 복합체라는 개념을 다시 건물에 실현한 경우다. 강렬한 사선들과 작은 스케일로 만나는 선과 면의 접촉이 도시의 연결과 또한 어긋남까지 담은 듯하다. 휴먼 스케일의 연결과 어긋남의 편린들이 반복되어 추상화된 표면을 이루고 있다.

이 작품 이후 다소 투박한 매스와 구조체들이 강조된 건물들은 이런 접근이 변이 과정을 겪은 복합체의 한 종(種)임을 보여준다. 추후에 다면체적 매스로 구체화된 다양한 버전의 건물들은 예화랑과 성동문화복지센터의 유전자를 변이한 형국이다. 하나의 모티브를 포착하여 일생 동안 변모를 지속해온 화가의 작품으로, 모네의 수련과 김창열의 물방울 그림을 예로 들 수 있다. 물방울 그림이 다양한 배경의 캔버스 위에 그려지며 변이를 지속하듯이, 운생동의 복합체도 건축 프로젝트의 상황에 맞게 적용되며 변이한다.

최근 운생동의 작업은 클라이언트에 따라 세 갈래로 나뉘어서 진행되는데, 이 점이 건축의 유형에 그대로 드러난다. 이는 공공 기관에서 발주하는 설계공모 작업, 개인 건축주 프로젝트, 시행사 주도 프로젝트로서, 사실 잘 운영되는

Gallery Yeh (2015) is the signal work that raised the profile of UNSANGDONG to one of the country's leading architects. Liberally edged-out concrete exterior coverings named the 'urban canvas' occupied the background to the intricately-positioned buildings and their surroundings, inserting holding gaps and spaces between sections. The singular quality is that, similar to Kim Incheurl's minimalistic KimOkGill Memorial Hall, the work takes on a 'pure abstract' form in its edged-out concrete walls, as a key example of a work of urban architecture from the 2000s. If the original exposed concrete walls had a vertically-standing appearance, Gallery Yeh, with its liberal corners, follows its own natural proportions with a certain indifference. It is not an overstatement to say that UNSANGDONG, employing techniques to create visual impact through irregular geometry, was influenced by Kwanseok Kim of Artech Architects in terms of its formative style . The architecture of Kwanseok Kim of the 1990s, while impactful in his emphasis of an autonomously edged-out geometry in the façade and form through the use of surfaces and masses, remained as something 'accessible', and other so-called 'mass-produced residential buildings' could easily copy this approach. In spite of his quest to free the façade, which brought about a change in his way of thinking, it did not receive the full depth of his thought. In contrast, however, by bringing out the compositional changes to emancipate the façade, Gallery Yeh functions as an urban canvas that is not easily reproduced. It shows how compositional surplus brings about a freedom for the surface.

If Gallery Yeh emphasized how it differs from the situation found in the neighboring city, Seongdong Cultural & Welfare Center (2013), sufficed as a description of an image of an open public building. The movements of the interior staircase and the diagonal lines of the composition cleverly created an open and virtual urban image. Fitting the theoretical concept of a Compound Body as suggested by UNSANGDONG, the architecture brought about a state in which the movement of people can be well-observed through the interior and the exterior of the building. It is a case where the concept of Compound Body, which was derived from a careful study of the urban phenomenon, is realised in architecture. The powerful diagonal lines and the small-scale contact between lines and surfaces seem to even embody the sense of connectedness and disorder within the city. Through this repetition of connected and the disharmonious partitions on a human scale, an abstract surface is created.

The works with emphasis placed on the somewhat crude masses and structures that followed reveal themselves to be under the same genetic family as Compound Bodies, but subject to certain modifications. The various structures that manifest with polyhedral masses in the later years come from a genetic modification of Gallery Yeh and Seongdong Cultural & Welfare Center. One might be reminded of Monet's Water Lilies or Tschang-Yeul Kim's Water drops as examples of a series of paintings that offer variations on an early original motif. Just as the painting of

설계사무소가 클라이언트층을 확보하고 있는 상황과 크게 다르지 않다. 또한 많은 설계공모에서 상위권을 유지하는, 운생동의 작업 유형도 몇 가지가 있듯이, 나머지 프로젝트들도 몇 가지 유형으로 정리할 수 있다. 설계공모에서는 보다 개념적이며 논리적인 작업 유형이 있으며, 나머지 시공을 전제로 한 프로젝트들은 보다 감각적이며 직설적인 유형에 속한다.

본 비평에서 본격적으로 다룰 강남의 프로젝트들은 각기 다른 컨텍스트에 위치하여 각각 다른 반응이 건물의 표면에 구축되어 있다. 강남에도 다양한 겹의 오피스 건축과 도시주거 건축의 유형이 있듯이, 주변의 미시적인 컨텍스트에 즉각 반응하는 경향이 보인다. 청담동에 위치한 '몽유도원도 이상봉타워(이하 이상봉타워)'에서는 '크링'(2008)에 적용된 원형의 모티브가 보인다. 조각이냐 건물이냐 하는 논쟁을 불러일으켰던 크링은 원형의 패턴이 입체화된 모습을 지녔고, 대로에 면하여 아주 큰 패턴의 입면을 만들어 시선을 끌고 있다. 크링에서 구사된 텍토닉은 논리적으로 이해되기보다는 김창열의 물방울 그림과 같은 입체 감각을 건물에 투사한 격이다. 입면의 폭과 높이에 무관하게 원형의 패턴은 완전하거나 잘린 채로 표현되어 도시추상을 보이고 있다.

이상봉타워도 대로에 위치하며 많은 해외 건축가들이 설계한 패션 업체의 사옥들 사이에 있다. 바로 옆의 올슨 쿤디그(Olson Kundig)의 큰 건물이 뉴욕에 있는 미스의 시그램 빌딩과 같은 매트한 검정색으로 대로에 면하여 아우라를 지니는 모습이라면, 이상봉타워는 비교적 좁은 폭의 입면을 지니며 원호가 어우러진 수직 루버들의 모습으로 한글을 강조하는 이상봉 디자이너의 태도처럼, 해외 건축가들의 건물들 사이에서 존재감을 유지하고 있다. 루버는 '건축의 겉'으로 부차적으로 붙어 있어, 청담동의 건물들이 이용하는 이중 입면의 모습과 크게 다르지 않다. 무광의 느낌을 만드는 효과에 치중한 점은 크링에서 입면을 만드는, 감각적인 접근과 유사해 보인다. 다만 크링에서의 실험적인 면은 표현적인 미묘한 감각 형성을 위해 조절된 모습이다. 아파트 모델하우스로서 크링이 굉장히 자유로운 입면을 가진 것과 달리, 프로젝트 파이낸싱으로 진행돼 주거층과 오피스텔 등의 구성이 중요했던 이상봉타워는 복잡다단한 단면을 건물의 정면에서는 원호의 루버로 추상적으로 만들고, 옆면에서는 프로그램이 추측되는 입면을 만들었다. 패션산업과 도시주거가 결합된 복합체의 이미지로서, 각자의 아이덴티티가 부각되는 청담동이라는 컨텍스트에서 세장한 건물의 비례로 인하여 약간의 굴곡만 느껴지는 비교적 눈에 띄지 않는 수수한 형태의 입면이다.

이 책에서 소개되지는 않지만 최근작 '지엘타워'는 9호선이 개통되면서 오피스텔 등이 많이 들어선 봉은사로에 위치하고 있다. 무심한 듯한 박스와 유리 매스가 쌓인 형국으로 스케일이 분절된 덕택에 주변의 오피스텔에 비해 눈에 잘 띄지

the water drops are repeatedly drawn on various canvas backgrounds in different forms, the Compound Body of UNSANGDONG too adapts and changes according to the demands of the architectural project.

The recent works of UNSANGDONG can be divided into three separate pursuits, and this is clearly shown in the respective architectural types. This is not so different from the division supposed between an architectural planning competition project for a public institute, a project by an independent architect, and a project directed by a developer with a well-managed architectural office and a secure base of clients. Moreover, just as there are types of work from UNSANGDONG that usually get to the upper tier of many design competitions, the other remaining projects can also be classified into a certain number of types. For example, there is the kind of submission that is more conceptual and logical than the one sent to design competitions, while there is also the kind that is designed with construction in mind and is thus more intuitive and straightforward.

Having settled into each of these different contexts, the projects in Gangnam, Seoul—the subjects of this critique—have embraced different respective responses to their building surfaces. As there are multiple layers of office architecture and types of urban residential architecture in Gangnam, there is a tendency for projects to respond immediately to the most minute contextual variations in the surroundings. In Mongyudowondo Tower, which is located in Cheongdam-dong, a circular motif that was used in Kring (2008) can be found. Kring, which once stirred up a debate regarding its definition as a either sculpture or a work of architecture, wears a virtualized appearance of a circular pattern, and attracts viewers by attaching a large façade of the pattern next to the main road. The tectonic used in Kring, similar to Tschang-Yeul Kim's Water drops, projects a sense of virtuality onto the architecture that deemphasizes logical understanding. The circular pattern, in no relation to the width and the height of the façade, is expressed in either a perfect or an incomplete state, and contributes itself to the theme of urban abstraction.

Located along the main road, Mongyudowondo Tower is also positioned in the midst of fashion house buildings designed by international architects. While the large adjacent structure by Olson Kundig seems to effuse a certain aura by being dressed in matte black and situated next to the road, Mongyudowondo Tower has a relatively narrower width, and enforces its existence amongst the internationally-designed architectures with vertical louvers that are embraced by circular arcs. Being only secondarily placed as part of 'the architecture exterior', the louvers are not much different from the double skin that has been incorporated by many buildings in Cheongdam-dong. Also, the sensual focus of making the louver non-reflective seems to resemble the approach used to create the façade for Kring, although in this case it also seems that the experimental attitude of Kring was mildly adjusted to create an expressive and subtle sensation. If it may be said that

않는다. 오피스텔의 반복적인 입면이 무성한 봉은사로에서 다섯 개의 거친 콘크리트 박스가 적층된 이미지를 통해 도시에 담담한 도시추상을 제시하고 있다. '도시추상'이란 어떤 프로그램의 건물인지 쉽게 알 수 없게 매스와 표면의 작용만을 추구한 것으로, 볼륨과 표면의 단순한 관계만 표현되어 건물 내부의 사용이 쉽게 추측되지 않으며 건물이 마치 도시의 배경처럼 서 있는 것을 의미한다. 이러한 건물의 표현을 두고 바람직한 태도인지 아닌지 묻기 보다는, 프로젝트의 성격과 자본의 구조하에서 건축가가 표현할 수 있는 운신의 폭이 정해진 경우가 많아서 건물의 형태와 입면이 담담히 결정된 경우가 많다는 점을 지적하고 싶다. 중요한 것은 건축가가 읽은 컨텍스트와 그 속에서 본인이 표현하는 방향성이다. 난무하는 개발 속에서 어떤 장면을 연출할 것인지의 문제를 마주한 운생동의 선택은 무심한 듯한 도시추상이라 할 수 있다.

미동전자 사옥을 리모델링한 작업, '퓨처리즘그리드 미동전자'도 도시추상에서 크게 벗어나진 않는다. 성동복지문화센터에서 사용했던 사선이 여기서는 그저 표면과 그 깊이의 효과로 차용되었다. 정방형 느낌의 입면 비례를 가지고 있는, 강남에 많이 있는 상업 건물 유형으로서 외관을 사선의 세라믹 박판으로 마감했다. 창의 면은 후퇴하여 사선 구조가 더 돋보이는 형국으로 표피적인 디자인으로 치부될 수 있는 것을 최소화했다. 주변의 비슷한 건물들 사이에서 사선과 깊이는 더 돋보이며 눈에 띄긴 하지만 딱히 오피스 같지 않은 모습 때문에 공교롭게도 패션 업체들이 임대해 사용하고 있으며, 다소 긴장감 있는 도시추상의 배경으로 존재하고 있다.

운생동의 강남 건물들은 개인 건축주가 의뢰한 프로젝트라는 성격 때문인지 퓨처리즘그리드 미동전자를 제외하면, 예전 운생동의 건축 경향에 비추어볼 때, 보다 논리적인 모습이어서 쉽게 눈에 띄는 편은 아니다. 어느 정도 원숙하게 시원시원한 디테일의 처리를 보여주며, 표현상 강조해야 할 부분에 대해서는 실험적인 모습을 드러낸다. 그러나 실험도 예전처럼 과도하게 특이성을 보여주기보다는 반복적으로 사용되는 재료의 다른 디테일적 구축 방법에 대한 모색을 위주로 한다. 어찌 보면 앞으로의 운생동 작업의 포커스는 표면에 있기보다, 가로와 연결되는 저층부의 입체적 형성과 프로그램에 있을 것 같다는 느낌이 든다. 이는 최근의 공공건축의 행보에서 드러나는 점이다. 성동 문화복지센터와 최근 일련의 설계공모 작업들에서 입면보다 땅과 만나는 방법에 대해 집중하는 모습이 보인다. '한내 지혜의 숲'에서는 공원의 작은 도서관이 책꽂이부터 만들어진다는 발상으로 땅에서 비롯된 가구로부터 작은 공공도서관을 만들며, 상황의 포착에 의한 건축적 해결에 치중했다. 설계공모의 작업들에서는 거대한 데크로 새로운 지형을 만들거나, 테라스형 매스로 입체적 접근이 가능한 안들을 제안하고 있다. 전체적으로

Kring exhibits an extremely unrestricted façade for an apartment model house, Mongyudowondo Tower, which is composed of residential floors and studios due to project financing. It has a rather complicated composition, by which its front is expressed abstractly through the circular arcs and the louvers while its side adopts a façade that suggests a certain programme. As the image of a Compound Body that combines the fashion industry with an urban residence, and due to its narrowed-down building proportion for its locational context of Cheongdam-dong, it features a simple façade that exhibits only a slight curvature and is thus relatively less eye-catching.

While it has not been covered in this book the recent work GL Tower is currently situated at Bongeunsa-ro, where many studio apartments have moved in with the opening of the subway line 9. Because its scale is fragmented due to its excessive use of boxes and glass masses, the building is not as visible as the surrounding studio apartments. At Bongeunsa-ro, where the repetitive stylized façades of studio apartments can be found, the five crude concrete boxes suggest an emotionless urban abstraction through its layered effect. Seeking merely the application of masses and surfaces to obscure which programme the building belongs to, 'urban abstraction' refers to the expression of a simple relationship between volume and surface so that the buildings can be portrayed as part of the urban background without revealing what it holds inside. Instead of questioning whether it is right or wrong with regards to this means of expressing a building's character, I would point out that the form of the building and its façade is mostly decided nonchalantly, as the margin of expression that an architect can tap into is usually already fixed under the specific direction and composition of the project and capital. What is important, however, is the architect's interpretation of the context and the direction of expression. It may be said that this kind of indifferent urban abstraction is UNSANGDONG's answer to the question of what would be deemed appropriate in this state of rampant development.

The remodeled Midong Electronics & Telecommunication Headquarters, named Futurism Grid, does not stray too far from urban abstraction. The diagonal lines used in the Seongdong Cultural & Welfare Center are reused here merely to create an effect for the surface and its depth. Following the type of commercial buildings with a front-facing façade proportion, commonly found in Gangnam, the exterior of the headquarters was finished with diagonal ceramic plates. By retreating the glass window surface towards the back, the diagonal structure is highlighted, and this helped to make the design look less superficial. The diagonal lines and the depth stand out more when compared to the other similar-looking buildings in the background. However, because it does not look like an office building, the building is currently being rented by fashion companies while also retaining its identity as an urban abstract with certain design tensions.

Perhaps due to the nature of the project request

다수 시도된 유형들이 반복되어 시도되고 있어서 운생동의 시그니처를 인지할 수 있다. 이런 현상 속에서 의미 있으며 생동감이 있는 작품이 탄생하는 것도 사실이다.

　　　　도시추상은 끝없이 진행되는 패턴을 생산하는 것일 수 있다. 건축적 편린의 반복에 따른 패턴은 휴먼 스케일적 요소에서 시작하여 작으며 거대하지 않다. 문, 창, 그리고 계단 등등에서 편린이 시작되며, 그 어울림과 어긋남이 반복하며 패턴이 만들어진다. 이우환의 모노화나 김창열의 물방울 그림에서 회화 기법의 차별성과 만든 패턴의 차이가 의미를 만들어내듯이 끝없이 일상과 차별화되는 패턴과 배경은 생성이 가능하다. 다만 전문인이나 인구에 회자되기 위해선, 어떤 도시추상을 하는지가 중요할 것이다. 건축가들 간의 커뮤니케이션이 가능한 전문화된 실험인지, 소비자들에게 회자될 수 있는 상품인지의 경계에 서서 도시추상화는 다소 유보적으로 읽힐 매니페스토를 던지며 존재할 상태를 찾는 듯하다. 22세기에 남아 '나 지금까지 이런 얘기를 하고 있었소' 하고 단언할 수 있는 존재 방식을 취한 셈이다.

made by the private building owner, the buildings designed by UNSANGDONG in Gangnam—with the exception of Futurism Grid—are not that eye-catching when compared with the historic architectural style of UNSANGDONG. The buildings display a skilled and satisfying finish in their detailing, and exhibit experimental elements in places that need to be emphasised for expression. However, unlike in the past when the experiments displayed a clear sense of singularity, the experimental elements now focus mostly on finding alternative and detailed construction methods of repeatedly used materials. In a way, it feels as if the current focus of UNSANGDONG is not on the surface, but is in the virtual formation and the program of the lower half that is in connection with the streets. This is something revealed from its recent public architecture. Along with Seongdong Cultural & Welfare Center, the recent series of design competition works focus not so much on the façade but on the method of how the buildings meet with the ground. Applying the idea that a small library in a park is created from what resides on its book shelves, in UNSANGDONG's Hannae Forest of Wisdom, a small public library is built from ground level furniture while also placing focus on an architectural solution through capturing the moment. In the works gathered as part of the design competition, UNSANGDONG attempted to build a new terrain by creating an immense deck, and proposed other possible alternatives for virtual accessibility through masses in the form of terraces. As a whole, due to the repetition of such types in its attempts, a certain trademark in the design style of UNSANGDONG can be sensed. It is a fact that meaningful and lively works are born from phenomena such as these.

　　　　Perhaps urban abstraction is the product of an endlessly continuing pattern. This motif of repeated architectural partitions began with human-scale elements, and there is no necessity for such aspects to be large in size. These partitions begin with things such as doors, windows, and stairs, and a pattern emerges in the repetition of their harmonies and disharmonies. As the distinctiveness and the difference of patterns create meaning in Lee U-fan's Mono-ha and Tschang-Yeul Kim's Water drops, the production of patterns and backgrounds that differentiate themselves infinitely from the daily life are made possible. It is only a matter of what kind of urban abstraction is brought about for it to become a popular issue amongst professionals and the general public. Standing in the realm of professionalized experiment, that is only communicable to the architects and the realm of a product for general consumers, urban abstraction appears to be looking for a place of existence while tossing out its somewhat tentatively-written manifesto. It is as if it has adopted a mode of existence through which it can outlive into the 22nd century and declare 'I've been saying this all this time'.

OCEANUS GROUP Haeundae Office
오션어스 해운대 사옥

Jung-dong, Haeundae, Busan, Republic of Korea **Site Area:** 1,294.60 m² **Building Area:** 771.29 m² 2,996.36 m² **Architects:** Jang Yoon Gyoo, Shin Chang Hoon **Designer:** Kim Sung Min, Kang Seung Soop, Ko Young Dong, Jang Chul Min, Ko Eun Jin **Client:** Ocean Us **Structure:** Steel Reinforced Structure **Photographer:** Jaekyeong Kim(Model), Sergio Pirrone

이 프로젝트는 스킨과 구조를 통합하는 방식의 건축을 제안한다. 투명성을 통하여 자연의 형태에서 발췌한 다양한 단면 패턴을 가진 스킨을 기능을 담는 거대한 통과 같은 구조로 구성하였다. 파도의 패턴과 나뭇가지의 패턴이 겹쳐지는 프레임의 공간은 내외부의 풍경을 조절하고 경사지를 따라서 쌓여진 오피스의 기능과 문화기능들의 주변 자연과의 결합과 소통을 생성하는 그릇과 같은 장치로서 작용한다. 대지 전후면 단면 레벨차 15m안에 주차 및 수장고+갤러리+갤러리 순으로 공간을 쌓아 올라가기 시작했고 중간 층의 주차장을 기준으로 상부에 2개 층의 업무공간을 더 쌓아 올려 요구된 프로그램이 구성된다. 상대적으로 개방감이 낮은 갤러리는 하부에, 업무공간은 상부에 두어 적절한 기능별 건축환경을 고려해 배치했다. 각 층의 공간에는 다양한 형태의 테라스가 생성되도록 의도적으로 매스를 조절하여 기본 레이아웃을 완성했다.

The scheme proposes an architecture that combines skin and structure. Through transparency, its exterior skin with varied profile patterns inspired by shapes in nature appears as a huge container embracing functions. A space framed with overlapping patterns of sea waves and wood branches controls landscapes of inside and outside. And this works as a bowl initiating integration and engagement of the surrounding nature and the office building's cultural and functional programs stacked up along a slope. Considering the site's 15m level difference between its front and rear sides, spaces are piled up in the order of a parking and storage, a gallery and another gallery. Above the parking space on the middle floor, two additional office floors are mounted to accommodate all required programs. Galleries with a relatively weak sense of openness are placed on the lower floors whereas offices are positioned on the upper floors to provide an appropriate architectural environment corresponding with a given function. In terms of basic layout, masses are strategically organized to give a different type of terrace to each floor.

Kolong E + Green House

코오롱 E + Green House

Location: Jeondae-ri, Pogok-eub, Cheoin-gu, Gyeonggi-do, Republic of Korea **Site Area:** 5,525 m² **Building Area:** 957.40 m² **Total Floor Area:** 1,837.04 m² **Architects:** Jang Yoon Gyoo, Shin Chang Hoon **Designer:** Kim Yoon Soo, Choi Young Eun, Ahn Hye Joon, Kim Ho Jin, Seo Hye Lim, Kim Mi Jung, Kim Ji Hye **Client:** Kolon Global Corporation **Structure:** RC Structure **Photographer:** Jaekyeong Kim(Model), Sergio Pirrone

산업화와 정보화에 의해서 진행되는 우리의 문명은 자연과 함께 공존해야만 하는 시대적 절대성에 직면해 있다. 건축을 통해서 행하여진 모든 문명의 흔적은 환경의 입장에서는 폐허를 의미하기도 한다. 미래적인 요구에 의해 문명이라 일컬어지는 것들조차도 이제는 죽어버린 건축적 장소와 공간을 만들어 나간다. 사회가 더욱 문명화 될수록 자연적 환경은 폐허를 향해서 나아간다. 자본주의의 보편화에 따른 소비지향적인 사회가 초래한 지구온난화, 자연고갈, 에너지고갈 등의 위험성이 더욱 대두되면서 환경에 대한 관심은 피할 수 없는 주제가 되었다. 특히 대도시의 인구, 건축, 인프라 등의 끝없는 팽창과 소모는 도시민의 거주공간의 새로운 전형의 요구를 피할 수 없게 되었다. 랜드스케이프 건축(Landscape Architecture), 에코 건축(Ecological Architecture) 등의 개념을 통해서 인공과 자연의 조합과 공생을 통해서 획득되는 친환경적 시스템의 요구도 여기에 있다.

Our civilization, having been progressed on industrialization and information, confronts to absolute demand for coexistence with nature. All the traces of civilization from architectural activities could mean ruins from the nature's point of view. What we call civilization is producing dead architectural space and places by futuristic demands. As the society is civilizes the natural environment goes towards ruins. As capitalism is on progresses, the consumption-oriented society makes focus on environmental issues making global warming, exhaustion of natural resource and energy draining a subject that we can't avoid. Especially the population of big cities, architecture, infra… because of theses endless expansions and consumptions, a new type of urban residence is required. Based on Landscape Architecture and Ecological Architecture, the demands for eco friendly system achieved by the combination artificiality and nature and symbiosis is also in that issue.

Primitive Boldness
원초적인 과감함

Keehyun Ahn (School of Architecture. Hanyang university)
안기현 (한양대학교 건축학부)

'어떤 것이 중요하지?, 무엇으로 시작하지'라는 질문으로 건축은 시작한다. 프로젝트마다 다른 대지 상황, 프로그램, 건축주의 요구 사항 등 무수한 변수는 다양한 해답을 가능하게 한다. 반대로 다양한 해답을 도출하는 한 건축가는 이러한 요인들에 영향받지 않는 스스로의 고유한 언어를 찾으려 한다. 스스로의 지향점을 명확히 가지고 있어야. 그가 건축을 통해 지속적으로 소통할 수 있을 것이다. 마치 망망대해 한복판에서 항해하는 배에 나침반 같은 것 말이다. 이런 관점으로 운생동의 나침반에 대해 생각해보고, 지나온 경로중 몇몇 지점들을 주목해보고자 한다.

Architecture begins with the questions: What is important? and How does it begin? Numerous variables, such as different site conditions, programs, and client requirements, enable various solutions to each project. Conversely, an architect who produces various answers seeks his or her own language that is not affected by these factors. Having one's own point of view clearly enables the architect to communicate continuously through their architecture. It is like a compass on a ship that sails in the middle of the sea. From this point of view, I would like to think about the UnSangDong's compass and focus on some points of the past.

01. 나침반. 복합체

'완전히 그 경계를 넘어서서 통합된 영역이 될 수 도 있을지 모른다. 건축과 타영역 간의 구분이 없이 완전한 <통합체>를 구성하기를 원한다. 태초에 하나의 세계였다. 잡음을 통해서 혼돈을 거듭하며, 본질음은 분산되었을 것이다. 우리는 시간을 거슬러 올라 서로의 영역을 통합하면서, 어쩌면 본질에 가까워질 수 있다. 복합체를 통해서 말하려는 것은 무수한 고리의 변형을 통해서 도달될 수 있는 경계선을 찾아내는데 있다. […]도시구조의 경계, 영여간의 경계, 타자와의 경계, 해석의 경계, 이해의 경계... 이 모든 경계를

01. Compass : Compound body

"It might be possible to achieve a complete integration beyond the boundaries. Wanted here is complete "Integration" devoid of distinction between architecture and other territories. Our beginning was one world. Through noise and chaos, its essence must have been dispersed. Suppose we can re-trace our times as we keep integrating each other's territories, we might be close enough to the essence again.   The intention behind discussing the compound body and countless transformation of its linkages is to discover a hidden boundary condition we can

모호하게 설정한다.[…], <찾기>를 통하여 내가 지각하지 못한 세계, 가능세계, 타영역을 매개로 하여 건축에게 주어지지 않은 세계를 발견할 수도 있다. 장윤규, 복합체, 간향미디어랩. pp31~32 참조

2005년 발간된 [복합체]라는 책을 통해, 운생동이 '어떤 건축을 하겠다'를 서술한다. 복합체는 서로 다른 성질의 것이 소통하고 교류하는 구조로서, 독창성을 만들어내는 플랫폼(platform)이 된다고 한다. 이 복합체는 목표점을 도달하기 위한 운생동의 이론이라고 말한다.

일반적으로 건축은 통일, 조화, 비례, 대칭, 리듬 같은 원리에 의거하여 여러가지 요소들(재료, 색체, 기하학, 구법, 마감재, 구조체, 빛, 동선 등)의 균형을 맞춰 나간다. 이 균형들이 세련미가 되어 우리에게 인식되는 반면 운생동의 작업이 균형을 이뤄내는 방식은 다른것 같다. 변화와 움직임 생생하게 교류하는 상태(마치 '기운생동'처럼)를 보여주 것 같다. 복합체라는 개념하에 서로 다른 성질의 것을 이용하기에 우리가 경험해 온 균형과 다르게, 과감하고 익숙하지 않은 풍경을 연출하는 것이 아닐까? 서로 다른 것들을 교류하게 하기 위한 목적을 추구하다보니, 일관된 구성과 짜임의 원리가 발견되기 보다 자유분방하고 역동적인 부분이 드러나는 것이 아닐까? 필자에게는 그 결과물들이 과감함을 넘어서 마치 날 것의 원시적인 모습을 보여주는 것 같이 느껴진다.

02. 경로 1. 인간이 동물되기 Becoming Animal

<인간이 동물되기>의 근본적인 물음의 시작은 본질 즉 아우라에 대한 물음과 연결되어 있다. 본질은 변화되지 않는 것으로 인식하는 것에 대한 강력한 발발이며, 본질은 고정된 것이 아니라 움직이는 것이라 정의하는 것과도 같다. 인간의 본질도 변화되는 것이며, 도시와 건축의 본질도,[…] 본질의 변신은 <동물되기>가 이끌어 내는 <강렬하게 되기>를 가능하게 하고 있다. <인간의 동물되기>의 설정은 인간과 동물이 만들어내는 잡종의 코드를 가지고 있다. […] 어떠한 한 분야만의 경계는 허물어지고 서로 상호간의 영향을 발휘하면 전체가 하나의 틀로 순환되기를 요구한다. 장윤규, 복합체, 간향미디어랩. Pp118 참조

동심원이 반복하며 만든 지형은 수직적인 파사드를 만들고, 이 수직적 지형은 원통의 브릿지로 연장되며 내부를 연결한다. 파사드가 된 프레임은 공간으로 변형되고, 그 변형된 공간은 브랜드와 연동하여 문화적 공간으로 탄생된다. 강렬하다. 운생동의 대표작중 하나인 크링이다.

새로운 모티브를 찾고자 하는 운생동의 집중은 곳곳의 프로젝트에 보여진다. [Motion Imagination] 파빌리온 프로젝트에서도 다양한 움직임을 만들수 있는 단면을 겹치고

reach. Boundaries between city structure, territories, people, interpretations, understanding,….are set to work ambiguously. The notion of <Searching> would allow us to recognize things not yet known, possible, and in mediation with other fields of territories, and thus we may confront the worlds yet to be offered to architecture."
Refer to Compound Body, Yoon Gyoo Jang, Ganyang Media Lab, pp. 31~32

The book Compound Body, published in 2005, it describes the "kind of architecture UnSangdong does." The compound body is a structure that communicates and interacts with others; it is said to be a platform for creating originality. This compound body is the theory UnSangDong uses to reach its target point.

In general, architecture balances various elements (materials, colors, geometries, structural methods, finishing materials, structures, light, movement, and etc.) based on principles such as unity, harmony, proportion, symmetry, and rhythm. While these balances are sophisticated and perceivable, the way in which the work of UnSangDong finds balance is different. It seems to show a change, movement, and astate of vivid interchange (as if it is Giunsandong). Unlike the balance we experience when using different qualities, doesn't the concept of the compound body produce a bold and unfamiliar landscape? If we pursue the purpose of interacting with each other, couldn't it be that a free-spirited and dynamic aspect is revealed, rather than principles of coherent composition and structure being discovered? It seems to me that the results are beyond boldness, showing the primitiveness.

02. Route 1: Becoming Animal

"The critical beginning of <Becoming Animal> connects with essence of aura. Essence is regarded as a fierce resistance towards thinking that it is not immutable. Essence is similar to a mobile thing rather than something transfixed. Human essence also transforms, like the ones of city and architecture, as well as our cultural and social essence. Transformation of essence potentiates <becoming intense> guided by <becoming animal>. <Becoming Animal> possesses hybridized codes generated by both a man and an animal. If the boundaries surrounding the disciplines collapse and the disciplines exchange influences reciprocally, all of them become part of one loop that circulates as a whole."
Refer to Compound Body, Yoon Gyoo Jang, Ganyang Media Lab, pp.118

The topography created by repeated concentric circles creates a vertical facade that is extended into the cylinder bridge and connects to the interior. The frame that becomes the façade is transformed into a space, and that transformed space is created to be a cultural place in cooperation with the brand. It is an intense Kring, one of the representative works of UnSangDong.

Their efforts to find a new motif can be seen in various other projects. The "Motion Imagination" pavilion

연결하여 과감한 형태를 만들어낸다. 움직임이 단면이 되고, 단면을 연결하여 날것의 형태를 만들고, 그 형태는 특별한 공간들이 된다. [Water Circle]과 [Life & Power Press] 건물은 각기 다른 기하학에서 수평적인 레이어들이 지형이 되고, 이 지형은 특별한 공간이 된다. 위에서부터 아래로 연결되는 3차원의 공간감은 극대화된다.

형태를 만들기 위한 여러가지 2차원적 모티브는 여러개의 z값을 통해 3차원으로 변환되고, 이 형태의 변환이 공간의 의미를 규정하는 단서가 된다. 이런 과정은 [S-Void building] 에서도, 최근작인 [한내 지혜의 숲]에서도 비슷한 양상으로 보여진다. 하지만, 대형설계사무소와의 협업때문인지, 프로젝트의 상황때문인지, 앞선 프로젝트들의 과감함이 발견되기 보다 세련됨이 더 드러나는 것 같아 아쉬움이 남는다.

그럼에도 불구하고 감각적/ 형태적 탐구를 통해 개발된 공간과 의미(동물되기)는 새로운 사유방식과 이미지를 생산해내는 것이 확실하다. 이것은 공간의 켜(Layer)들과 그 켜들에 투영된 프로그램의 조합으로 운생동의 하나의 유형이 되가고 있는 듯하다.

03. 경로 2. 클립시티 CLIP CITY

도시와 건축의 탐구에서 물리적 대상 자체에 관심을 가지기 보다는 그들이 만들어내는 보이지 않는 관계에 주목하고 있다. 대상과 대상 사이에 존재하는 그 무엇에서 도시와 건축의 본질을 발견할 수 있다고 보기 때문이다. 도시속에 존재하는 수많은 관계, 즉 순응, 대립, 공존, 자생 등은 그 자체가 개념덩어리이다. [...] 도시적 흐름은 무형의 개념이 유형의 건축으로 이르게 하는 관계의 경로이다. 과거에 물, 공기와 같은 자연의 흐름이 도시와 건축에서 중요한 주제의 하나였지만, 이제 새로운 개념의 흐름이 몰려오고 있다. 이러한 흐름에 대응하기 위해 건축은 더 이상 독자적인 개체가 아닌 관계를 만들어나가는 주체가 되어야 한다.
장윤규, 복합체, 간향미디어랩. Pp76 참조

독자적인 개체의 개념들을 운생동의 어휘로 정의하고, 그 정의된 개념들을 새롭게 관계맺게 한다는 것으로 보인다. [Culture Forest]는 우리가 익히 알고 있던 공적영역의 물리적, 정서적 변환시킨다. 이를 통해 포디움, 표피, 프로그램으로 진화되어 관계를 완성한다. 여러 방향의 사선들은 저층부를 거칠고 과감하게 공적영역으로서 드러낸다. 마치 가위로 난도질 당한듯한 포디움은 과감을 벗어나 과격해까지 보이지만, 그로서 자신의 역전된 관계를 소리쳐 알려주고 있는 듯 하다.

[White Cube Matrix]는 다른 척도(Scale)를 가진 아이들의 눈높이를 기준으로 새로운 관계를 만들어낸다. 서로다른 크기의 창을 가진 박스들이 자연, 문화, 예술, 사람들과 관계매즌 미디어가 되어 아이들의 상상력을 자극한다. 큐브들의 3차원적

project creates bold shapes by overlapping and connecting cross-sections to initiate various movements. The movements become a cross-section and create a raw shape by connecting other cross-sections; the shape becomes a special space. The "Water Circle" and "Life and Power Press" buildings in different geometries become horizontal terrains that come to this special space. The three-dimensional spatial sense is maximized from top to bottom.

Various two-dimensional motifs for making shapes are transformed into three dimensions through several z-values, and the transformation of this form is a clue that defines the meaning of space. A similar process can be found in the "S-Void Building," and "Hannae Forest of Wisdom." However, these seem to be more sophisticated than the boldness of previous projects, because of the collaboration with a large design office and the projects' situation.

Nevertheless, it is certain that space and meaning (being animals) developed through sensuous and morphological exploration produce new ways of thinking and images. It seems that this is becoming a UnSangDong, a combination of the layers of space and the program projected on these layers.

03. Route 2: CLIP CITY

"We focus invisibleness within the relationships among materials and substances, rather than their physical configuration. This is because we believe we can capture the essence of city and architecture out of those relationships. Various relationships such as-adaptation, confrontation, coexistence, self-organization, are concepts themselves. Urban flow corresponds to the path of relationships in guiding formless concept towards formal architecture. While themes such as water, air, and other natural elements were mainstreams of arranging urban flows in the past, the new conceptual elements of urban flows are discussed today. In order to respond to the tendency, architecture must not remain the object independent of its own, but the subject aggressive enough to create various relationships."
Refer to Compound Body, Yoon Gyoo Jang. Ganyang Media Lab, pp.76

The concepts of independent entities are defined as lexical vocabularies, and the newly defined concepts are related. "Culture Forest" transforms the physical and emotional aspects of the public domain we know. Through this, the podium, skin, and programs evolve to complete the relationship. Diagonals in various directions are exposed as rough and boldly public areas on the lower floors. The podium, which seems to have been smashed with scissors, appears bold and intense, but it reminds us of its reversed relationship.

"White Cube Matrix" creates a new relationship based on the eye level of children and along different scales. Boxes with windows of different sizes stimulate children's imagination by becoming media related to nature, culture, art, and people. Through a three-dimensional matrix

매트릭스를 통해 교실, 유치원의 새로운 공간적 가능성을 제시한다. 부분과 전체, 부분과 부분이 관계를 새로이하며 서로 복합화된다. 복합화된 관계는 새로운 이미지로 구현되고 있다. 여기서도 이 새로운 이미지는 아주 직접적이고 원시적이다.

04. 경로 3. Mythological Imagination
<Mythological Body> transposes imagination into reality and promotes architectural space and materialization. The space of mythological imagination does not just remove logos and fill the space with absurd stories.[···] Certainly a high level of development is needed in order to embrace both the imagination and its actual materialization. Then the imaginative city and its architectural proposition will not be left at the level of ridicule but instead there will be boundless possibilities for revitalizing the city and regenerating it into something new. Compound body, l'ARCK, pp 179

복합체를 이야기할 때 장윤규가 가장 먼저 언급했던 오프페우스의 시선이 떠오른다. 운생동의 건축은 실제적인 것들의 관계/변환에서만이 아니라 가상의 비물질적인 것들을 물질화하고 공간화하면서 새로운 건축을 시도한다.

[Ocean Imagination]과 [Communi-Imgination]의 두 파빌리온 프로젝트에서는 자연적 요소들이 중력이 반하며 건물에 쉽게 일체화되어 초자연적인 모습으로 제안되었다. 역시 세련됨으로 무장하기 보다 거칠고 과감하게 아이디어가 표현되고 있다. [Interactive Culture Stage]에서는 도시의 거대한 블록을 아무런 저항이 없는듯, 반전시키고, 조형화, 그리고 건물화 시키며 새로운 스케일의 공간으로 만들고 있다. 최근작인 [몽유도원도 이상봉타워]도 가상의 그림과 그림의 의미를 건물의 표피에 대입하여 도시맥락에 새로운 표정을 입혀냈다. 비록 가상의 것이 공간 전체에 통합되어 있기보다 표피에 머물고 있어보이는 아쉬움은 있지만, 역시 과감한 표피는 사람들에게 색다른 경험을 제공할 것은 의심의 여지가 없다.

00. 제3자의 기대
복합체를 토대로 한 운생동의 작업은 위에 언급한 3개의 경로이외에도 다양한 방법이 있을 것이다. 40이 되기전에 복합체라는 텍스트를 스스로 구축하고 실험해온 다양한 작업들은 필자에게 탐구하는 즐거움 이상의 자극이 되었다. 레이어들의 복합과 통합, 틈새만들기, 프레임만들기, 지형만들기, 경계지점을 발견하고 정의하는 일, 다양한 분야와 끊임없이 협업하고 관계맺기, 더불어 물질적인 건축만이 아닌 사유의 텍스트를 통해 구축하는 말하는 건축까지… 매력적이다.

of cubes, it presents new spatial possibilities for classrooms and kindergartens. The relationships between part and whole, and part and part, are newly combined. The complex relationship is implemented as a new image. Again, this new image is very direct and primitive.

04. Route 3: Mythological Imagination
<Mythological Body> transposes imagination into reality and promotes architectural space and materialization. The space of mythological imagination does not just remove logos and fill the space with absurd stories.[...] Certainly a high level of development is needed in order to embrace both the imagination and its actual materialization. Then the imaginative city and its architectural proposition will not be left at the level of ridicule but instead there will be boundless possibilities for revitalizing the city and regenerating it into something new.
Compound body, l'ARCK, pp 179

When I talk about the compound body, I recall a view of Orpheus, first mentioned by Yoon Gyoo Jang. UnSangDong's architecture tries to construct something new by materializing and spatializing not only the relationship or transformation of actual things, but also virtual, immaterial things.

In the two pavilion projects of "Ocean Imagination" and "Communi-Imagination," the natural elements are proposed as supernatural, and easily integrated into the building against gravity. Rather than being armed with sophistication, the idea is expressed roughly and boldly. In the "Interactive Culture Stage," the huge blocks of the city are reversed, shaped, and built to a new scale of space, with no resistance. The recent work "Mongyudowondo, Lie Sangbong Tower" also imparts the meaning of virtual paintings and drawings into the skin of the building, so that it has a newly urban look. Although it is an understatement that a virtual thing stays on the skin rather than being integrated into the whole space, there is no doubt that the bold skin provides people with a different experience.

00. Expectations of a third party
In addition to the three routes mentioned above, there are a variety of other ways that UnSangDong's works are based on the compound body. The various tasks that UnSangDong structures experimented with the text of the compound body and inspire me to more than just the pleasure of exploration. UnSangDong has combined and integrated layers; created gaps, frames, and terrains; found and defined boundary points; constantly collaborated with and connected various fields, and built an architecture constructed not only through material, but also through the text of reason; all of this is very attractive.

Finally, I am reluctant to describe my impressions, but I think UnSangDong's strength is in the bold, radical, and primitive movement seen in their vocabulary. As you might

마지막으로, 개인적인 표현이기에 조심스럽지만, 그들의 어휘에 드러나는 과감하고 과격한, 원시적인 혹은 원초적인 움직임에 강한 힘이 있다고 생각된다. 구름을 보면서 양떼를 떠올리고 산위에 바위를 보면서 누군가의 얼굴을 떠올리듯이, 운생동의 과감하고 원시적인 건물들이 도시속에서 강력한 연상작용으로 새로움을 계속해서 만들어주길 기대한다.

recall a herd of sheep while looking at the clouds or think of someone's face while looking at rocks on a mountain, I hope that the bold and primitive buildings of UnSangDong will continue to create newness and powerful associations with the city.

Theater Contour (Namsadang)
안성 남사당공연장 창고

Location: Bogae-myeon, Anseong-si, Gyeonggi-do, Republic of Korea **Site area:** 8,264.5 m² **Building area:** 2,411m²
Gross floor area: 3,147.19m² **Building to land ratio:** 31.56% **Floor area ratio:** 36.51% **Building scope:** B1, 2F
Height: 20m **Photographer:** Jaekyeong Kim(Model)

한국의 땅, 한국의 초가집을 추상화 하여 남사당 공연장을 제안한다. 남사당 공연장을 통하여 한국의 문화, 한국의 놀이를 담을 수 있는 진정한 우리의 놀이 공간을 제공할 것이다.
남사당의 풍물, 어름, 살판, 덧뵈기, 버나, 덜미로 구성된 남사당의 여섯마당은 우리나라 최초의 대중예술의 형식이다. 여기서 남사당 공연장을 계획하는 방식으로 여섯마당의 다양하며, 인터랙티브공연을 담을 자연적이며, 한국적인 공연의 그릇을 제안한다. 여섯마당의 공연장은 단순한 내부 공연장이 아니라 외부의 자연과 연계되어지는 열려진 자연의 공연장을 제안한다.

A performance hall for Namsadang (a Korean folk band), inspired by the Korean land and traditional thatched houses, is proposed. Through the Namasadang performance hall, we will provide a true playground where Korean culture can be showcased and traditional plays can be shown.
The six acts of Namsadang, composed of Pungmul-nori (performance), Eoreum (Chosun tightrope dancing), Salpan(tumbling), Deotboigi (a mask dance), Beona (a play showing various skills revolving around an instrument played with a sharp-ended stick), and Delomi (a traditional puppet show) were the first forms of popular art in Korea. Here, as a way to plan the Namsadang performance venue, we suggest a natural and Korean style that will accommodate the six varied and interactive performances. The performance hall is proposed to be open venue linked to the nature outside, not just an enclosed performance hall.

Asian Culture Complex in Gwangju
(International Competition, 3rd prize)

광주 아시아 문화전당
(국제현상설계 3등안)

Location: Dong-gu, Gwangsan-dong, Gwangju, Jeonnam, Republic of Korea **Site Area:** 93,375 m² **Building Area:** 65,213 m² **Total Floor Area:** 146,672 m² **Architects:** Jang Yoon Gyoo, Shin Chang Hoon, Kim Woo Il, Kim Woo Young **Designer:** Yeon Kyong Hee, Kim Yoon Soo, Kim Seong Min, Choi Hye Jin, Kwon Woo Seok, Jung Bok Ju **Photographer:** Jaekyeong Kim(Model)

광주아시아문화전당을 통하여 문화를 담는 새로운 그릇과 같은 스테이지를 제안한다. 새로운 도시의 스테이지는 길과 광장을 관통하며 만들어내는 비워진 광장과도 같다. 이는 주어진 프로그램과 랜드스케이프를 관통하는 거대한 지도를 만들어 주는 작업이라 볼 수 있다. 생성과 소멸 사이의 대지로서의 아시아문화전당을 제안한다. 대지의 기억과 역사의 관계를 도시적이며 대지적인 지형의 조각으로 변환한다. 아시아 문화전당은 새로운 대지로서의 광장을 만드는 작업과 같다. 광장의 프로그램은 시간과 사람들에 의해 언제든지 변할 것을 수용하며 생성과 소멸의 역사를 기록하는 장소로 남는다.

We propose to constitute a stage that acts as a new container for the culture and everyday events. If new stage for the city is creating an empty open space out of penetrations pathways and plazas, making the new 'city stage' is as same as to conduct an enormous mapping operation with penetrating programs and landscape. We propose the Asian Culture Complex as the earth in between creation and disappearance. The earth that is transformed into fragmented pieces and become urban landscape filled with memories of the site and historical evolution of various spectrums of relationships.

 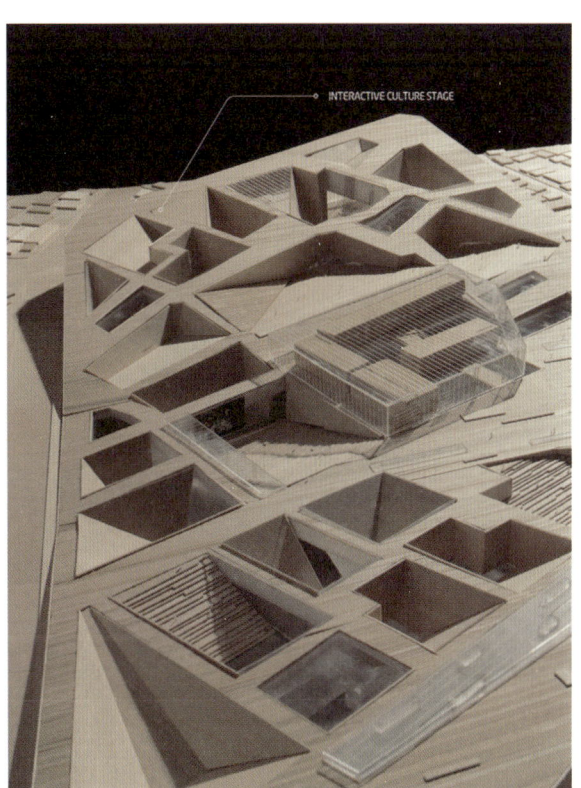

Sanghoon Youm (Department of Architectural Engineering at Yonsei University)
염상훈 (연세대학교 건축공학과)

현대 도시는 '체스(Chess)'보다는 '바둑(Go)'에 가깝다. 들뢰즈(Deluze)와 가타리(Guattari)는 1986년 [Nomadology: The War Machine]에서 체스와 바둑을 비교하고 있는데, 체스는 '전방, 후방과 전투(battle)가 존재하는 규제되고, 제도화되고, 암호화된' 전쟁인 반면, 바둑은 '전선(battle line)이 존재하지 않고 대치 및 후퇴가 없는, 순전히 전략만 존재하는 전쟁'이며 열린 공간에서 말의 배열에 따라 전세가 변하는 전략적인 게임이라고 설명하고 있다. 바둑은 체스와는 달리 말에 어떠한 가중치도 없고 순전히 말이 놓이는 위치가 만들어내는 관계에 따라 힘이 형성된다. 또한, 말이 놓이는 시작점과 근원이 없기 때문에 지엽적으로 다양한 힘의 관계가 생길 수 있다.

바둑처럼 기존의 바둑알 사이에 새로운 바둑알이 끊임없이 놓이고 소멸되면서 매번 바둑게임의 전세가 바뀌어가고 있는 형상은 현대 도시에서 끊임없이 발견된다. 이렇게 주변 문맥(context)이 빠른 속도로 변화하는 상황에서는 변화 그 자체를 건축에 수용하면서 건축과 도시를 바라볼 필요가 있다. 도시를 정지된 장소(rigid place)가 아니라 변화하는 상황(transforming situation)으로 바라볼 필요가 있으며 건축은 힘의 균형을 바꾸어가는 작용물질(agent)로 이해할 수 있는

The modern city is closer to "Go" than "Chess." Delueze and Guattari compared chess and Go in Nomadology: The War Machine in 1986. Chess is a regulated, institutionalized, encrypted war in which front and rear battles exist. On the other hand, Go is a strategic game, in which there is no battle line, there is no substitution or retreat, and only strategy exists. The war situation changes according to the arrangement of pieces in open space. Unlike chess, Go has no weight on pieces and power is created by the relationship of the position that the piece rests on. In addition, since there is no starting point and source of pieces, various power relations can be generated peripherally.

What the war situation of the Go game is constantly changing is a new piece is constantly placed and extinguished between the existing pieces. In the modern city, this concept is constantly found. In such a case where the surrounding context changes rapidly, it is necessary to view the architecture and the city while accommodating the change itself in the architecture. It is necessary to view the city as transforming, not rigid, and architecture can be understood as an agent that changes the balance of power. In other words, it not only responds to change but also looks at architectural design as a process that responds appropriately to the balance of power in the city.

것이다. 즉 변화에 대응할 뿐만 아니라 도시에서 힘의 균형에 적절히 대응해나가는 과정으로 건축 디자인을 바라보게 되는 것이다.

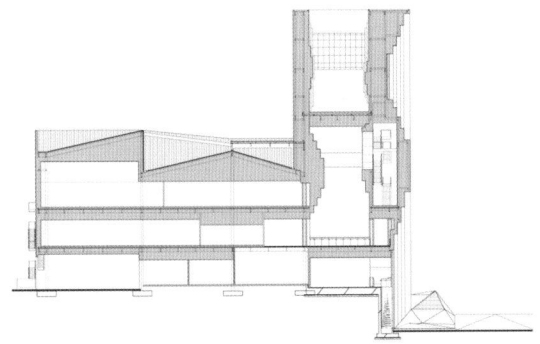

그림 1. 크링(Kring), 2006

이러한 관점에서 운생동건축사사무소(이하 운생동)의 건축 작업을 본다면 도시를 대하는 운생동만의 독특한 입장을 발견하게 된다. 운생동의 건축은 도시에 반응하여 순응하기보다는 도시에 영향을 주어 새로운 도시를 만들고자 하는 의도가 있다고 여겨지는데, 빈공간을 만들어 도시와 건축의 흐름을 연결하고자 하는 방식보다 도시와의 '경계의 형상'을 바꾸고 도시와 건축 사이를 이격시켜 도시를 새롭게 바라보게 하려는 시도가 강하게 드러난다.

건축과 도시, 건축과 조경, 천정과 바닥, 기둥과 벽 등의 여러 건축적 요소들간의 구분을 없애려는 운생동의 의도는 경계를 없애거나 모호하게 만들려는 것이 아니라 여러 요소들의 복합체로서 경계 자체가 독립된 개체로 존재하기 바라는 것처럼 보인다. 건축과 도시의 경계를 흔히 입면, 표면 혹은 외피라는 단어로 표현할 수도 있겠지만 운생동 건축의 경계는 – 운생동에서 '스킨스케이프(skinscape)'라는 독립된 단어를 사용하듯이 – 경계 자체를 독립된 개체로 이해할 필요가 있고 외피의 역할 또한 다른 건물의 그것과 다르게 이해할 필요가 있다. 운생동 건축의 외피는 건물을 둘러쌓고 있는 것도 아니고, 반대로 도시를 덮고 있는 것도 아니다. 도시와 건축 사이에 간격을 만들고 도시도 건축도 아닌 제 3의 공간을 만든다. 건축적인 관점에서 운생동 건축의 입면은 내부 공간의 성격과 관계를 반영하지 않고 있다고 보일 수 있겠지만, 운생동 건축의 입면은 도시와 건축을 연결시키는 공간이라기보다는 도시 일상으로부터 새로운 세계로의 탈출과 경험을 만들고자 하는 것으로 이해할 수 있을 것이다.

이러한 접근은 서울이라는 도시에서 더 큰 의미를 갖는다. 서울 도로의 폭 때문에 그러한데, 서울의 도로는 테헤란로와 같이 도로의 폭이 넓고 차선의 수가 많거나 주택가 골목과 같이 보차분리가 되어있지 않은 작은 이면도로인 경우가

From this point of view, if you look at the architectural works of the UnSangdong Architects' Office (UnSangDong), you will find a unique position of UnSangDong toward the city. It is believed that the architecture of UnSangDong is intended to make a new city and influence the city rather than adapting and responding to the city. However, the attempt to change "the shape of the boundary" within the city and to separate the city from the architecture, rather than connecting the city and the flow of architecture by creating an empty space, is strongly implemented.

The intention of UnSangDong to destroy the distinction between architecture and city, architecture and landscaping, ceiling and floor, columns and walls, is not to eliminate or obscure boundaries. It seems as if the boundary itself exists as a separate entity as a compound body of several elements. The boundaries between architecture and city can often be expressed by the word surface or skin. But the boundaries of the UnSangDong architecture need to be understood as an independent entity, just as the independent word 'skinscape' is used by UnSangDong. The role of the skin also needs to be understood differently from that of other buildings. The exterior skin of the UnSangDong architecture is neither surrounding the building nor covering the city. On the contrary, it creates a distance between the city and the architecture, and creates a third space, neither a city nor a building. From the architectural standpoint, it can be seen that the facade of UnSangDong architecture does not reflect the character and relationship of interior space. However, it can be understood that the facade of UnSangDong architecture is not a space connecting city and architecture but rather an attempt to create an escape from urban life and an experience of a new world.

Because of the width of Seoul's roads, this approach has more significance in Seoul. Seoul's roads are often small side streets not separated from each other like a residential alley, a wide road width, or a large number of lanes such as Teheran Road. This is very different from a situation where there are two or three lanes and a wide pedestrian walkway on one side, such as in a foreign city, especially New York. This is the reason why it is difficult to directly apply the foreign recreation method of pedestrian paths and Pocket Parks in Seoul. It is the private area such as a public site that is eventually plays an important role in the situation of Seoul roads or an architectural area such as the elevation or a low floor. Architecture is born with some degree of publicity, but the elevation of architecture in Seoul energizes the urban experience when it is a volume or space that is deeper than thinner.

많다. 외국, 특히 뉴욕과 같이 일방도로의 2~3차선 차로와 넓직한 보행도로가 있는 상황과는 다분히 다르다. 외국의 유수한 보행도로 재생 방법과 포켓공원 사례가 서울에 직접적으로 적용되기 어려운 부분도 이러한 이유이다. 이러한 서울 도로의 상황에서 결국 중요한 역할을 하게 되는 것은 공개 공지와 같은 사적 영역이며 입면 혹은 저층부와 같은 건축 영역이다. 건축은 태생적으로 어느 정도의 공공성을 타고나지만 서울에서 건축의 입면은 얇은 면일 때보다 깊이가 있는 볼륨 혹은 공간일 때 도시 경험에 활력을 넣을 공공적인 역할을 할 수 있게 된다.

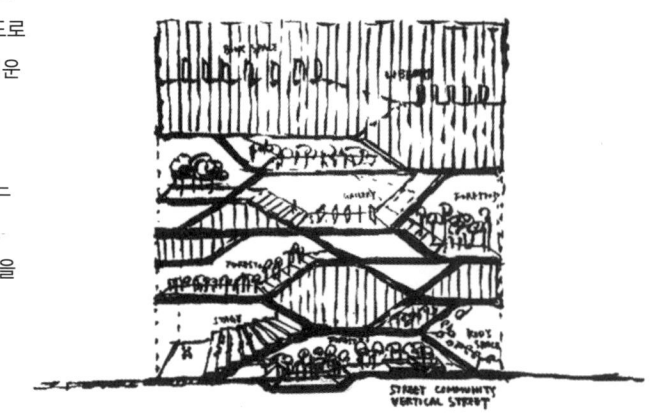

그림 2. 성수문화복지회관, 2013

운생동 건축의 경계는 실내와 실외를 구분 짓는 '면'이라기보다 두 영역 사이의 관입된 볼륨이고 이러한 특징 때문에 운생동 건축은 공공성을 담을 수 있는 잠재성을 갖고 있다. 운생동의 2006년작 '크링, 금호복합문화공간'(그림 1.)의 경우, 기존 모델하우스의 상업적 기능에 복합문화시설이라는 프로그램이 포함되어 기업의 문화적 이미지를 만들어 낸다. 기존 모델하우스 건물 앞에 추가된 두꺼운 볼륨은 건축과 도시, 상업성과 공공성 사이에서 새로운 가치를 만들고자 한다. 문화의 숲이라는 개념에서 출발한 '성수문화복지회관'(그림 2.)에서 가장 특징적인 부분은 새로운 지형을 만들고 있는 저층부이다. 사선의 기하학적 패턴을 활용하여 다양한 외부 공간과 계단 구조를 만들고 있다. 도시의 구조를 집약한 듯한 저층부 디자인은 내부 공간과 외부 외피를 동일하게 이해하고자 하는 접근으로 보이고 성수문화복지회관 개념 드로잉은 이러한 의도를 명확히 드러내고 있다. 저층부의 볼륨 자체가 입면이라고 봐도 과언이 아닐 것이다. 비실현작에서 가장 눈에 띄는 것은 광주 아시아문화전당 국제현상안이다(그림 3.). '문화적 캔버스'를 만들고자 하였던 운생동의 의도대로 이 캔버스 공간은 도시와 접하기도 하고 이격되기도 하면서 도시에 다양한 문화적 활동이 일어날 수 있는 플랫폼을 제공하고 있다. 이 건물은 배치도가 입면도라도 할 수 있을 정도로 수직과 수평의 관계가 복합적이고 이 설계안은 도시에 깊이감을 더하고 있다.

The boundaries of UnSangDong architecture are intrusive volumes between two areas rather than the "sides" that distinguish between indoor and outdoor, and because of these characteristics, UnSangDong architecture has the potential to contain publicness. In the case of UnSangDong's "Kring, Kumho Complex Cultural Space" created in 2006 (Fig. 1), the commercial functions of the existing model house include a program called a complex cultural facility, which creates a corporate cultural image. The thicker volume added in front of the existing model house adds new value between architecture, city, commercial, and public areas. The most unique characteristic of the "Seongsu Culture Complex" (Figure 2), which started from the concept of a cultural forest, is the lower layer that creates a new topography. A geometric pattern of diagonal lines is used to create various external spaces and stair structures. The low-rise design seems to be an approach to equally understand the inner space and outer skin. The drawing of the concept of the Seongsu Culture Complex clearly reveals this intention. It would not be an exaggeration to say that the volume of the lower floor itself is the elevation. The most prominent of the non-built works is the entry in the international design competition of the Gwangju Asian Culture Complex (Figure 3). The canvas space, which was intended by UnSangDong to create a "cultural canvas," was designed to provide a platform where diverse cultural activities could take place in the city in contact with the city and in the distance. This building has a combination of vertical and horizontal relationships so that the layout can be the elevation drawing, and this design adds depth to the city.

운생동의 건축이 이질적이고 특이한 것은 매스의 조형적인 측면이나 형태적인 측면이 아니라 도시와 건축 사이의 경계가 특별하기 때문이다. 주변 환경의 여러 요소들이 융합되고 변이되어 이질성이 짙은 형상을 띄게 되는데, 그런 이유로 운생동 건축의 경계 형상은 그 자체의 두께와 무게를 갖고 운생동이 추구하는 '복잡함과 현란함 속에 불편함과 불안함의 구조'가 가능하게 되는 것이다([SPACE(공간)] 609호 참고). 운생동 건축은 마치 바둑판 위의 체스말과 같다. 바둑의 규칙과 원칙을 따르지만 바둑이라는 세계를 뛰어넘는 세상을 꿈꾸고 있다. 바둑판의 바둑알이 게임의 균형을 알려주는 역할이라면 바둑판의 체스말은 바둑의 존재 이유에 대한 근본적인 질문을 끌어내고 있다. 운생동의 건축은 치열하게 바둑이라는 게임을 하면서 성장하고 있는 현대 도시에 잠깐 다른 생각을 하게하는 짧은 여행과도 같다.

UnSangDong architecture is unique and different because the boundary between the city and the architecture is special, not a formative aspect or an aspect of the mass. Various elements of the surrounding environment are fused and mutated to form a heterogeneous shape. For this reason, the boundaries of the UnSangDong constructions have their own thicknesses and weights. Therefore, it is possible to make a structure of discomfort and anxiety in the complexity and glamor pursued by UnSangDong. (See [SPACE] 609) The UnSangDong architecture is like a chess piece on a checker board. It follows the rules and principles of Go, but dreams a world beyond the world of Go. If the piece on the Go board tells the balance of the game, the chess piece on the Go board draws out a fundamental question about why Go exists. The architecture of UnSangDong is like a short trip that adds a different thought to the growing modern city while playing a game called Go.

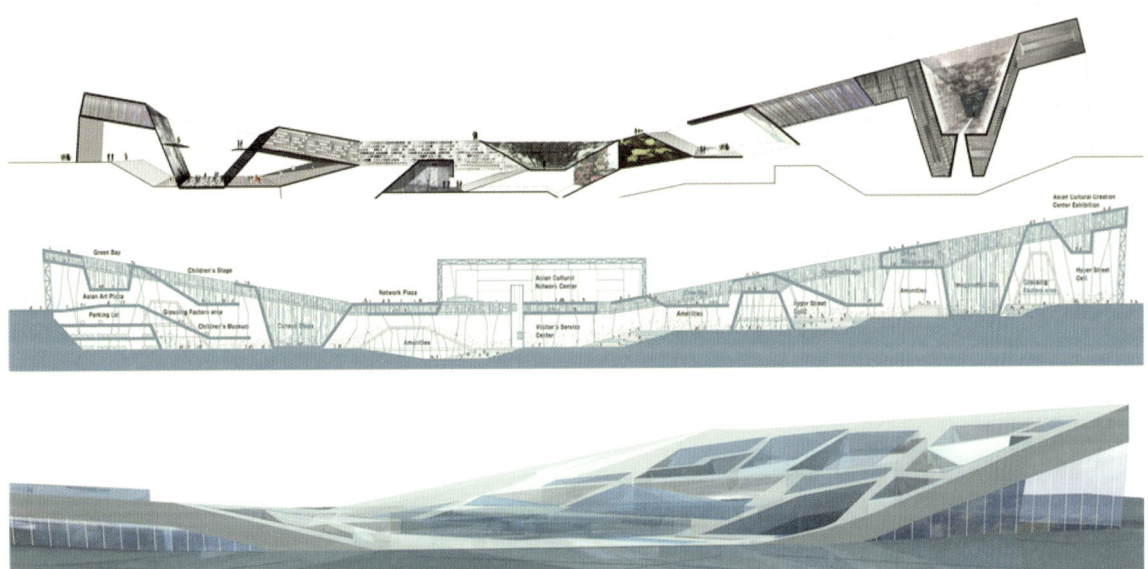

그림 3. 아시아 문화전당 국제현상 3등작, 2005

KTNG Culture Complex
(Competition Winner)

KTNG 복합문화센터

지상으로 부터 자유로운 비행 도시를 상상해보십시오. 이 떠다니는 도시는 신선함으로 가득 찬 공중 도시이다. 젊음은 상상력이 풍부한 정신으로 새로운 환상을 길러 낸다. 비행 도시로부터 거대한 상업 및 문화 공간, 수많은 도시 경관이 층층이 쌓여 있다. 이 무중력 구역은 거리, 광장, 숲, 상점, 놀이 공원 및 공원을 재배치 한다. 이것은 지구로부터 자유로운 상상의 도시이다. 다문화 공간으로서의 연극 공간은 비행 도시의 젊은 층에게 그들이 주도적인 역할을 할 수 있도록 적극적으로 도울 것이다. 비행 도시의 트레이드마크인 연극 공간은 쇼핑, 게임, 엔터테인먼트, 멀티미디어, 교육, 클리닉 및 커뮤니티 활동과 같은 이벤트를 구축하여 다양한 상상의 자원을 제공 할 수 있는 문화 센터 역할을 한다.

Imagine a flying city free from the ground. This floating city is an urban hovering in the air saturated with freshness, The youth mesmerized with this imaginative spirit nurture new fantasies, Numerous cityscapes are clustered together layer after layer and from the flying city of vast commercial and cultural space. This gravity-free zone rearranges streets, plazas, forest, shops, amusement park, and park. It is a city of imagination that is free from the earth. Theatrical Space as a multicultural space, the flying city will actively help young people to perform leading roles in expanding stage volume. The theatrical spaces, the trademark of the flying city, serves as a cultural center that can provide various imaginative resources by building events such as shopping, game, entertainment, multimedia, education, clinic, and community activities.

Seongdong Cultural & Welfare Center
성동문화복지센터

Location: 850-523, Seongsu-dong, Seongdong-gu, Seoul, Republic of Korea **Site Area:** 2,204 m² **Building Area:** 1,014.69 m² **Total Floor Area:** 9,558.75 m² **Architects:** Jang Yoon Gyoo, Shin Chang Hoon **Designer:** Kim Sung Min, Kim Min Tae, Seo Hye Lim, Ryu Sam Yeol, Ahn Hye Joon, Kim Won Il, Ahn Boo Young, Sim Jehyun, Kim Mi Jung, Jo Eun Jung **Client:** Municipality of Seongdong-gu **Structure:** Steel framed reinforcement concrete **Photographer:** Jaekyeong Kim(Model), Sergio Pirrone

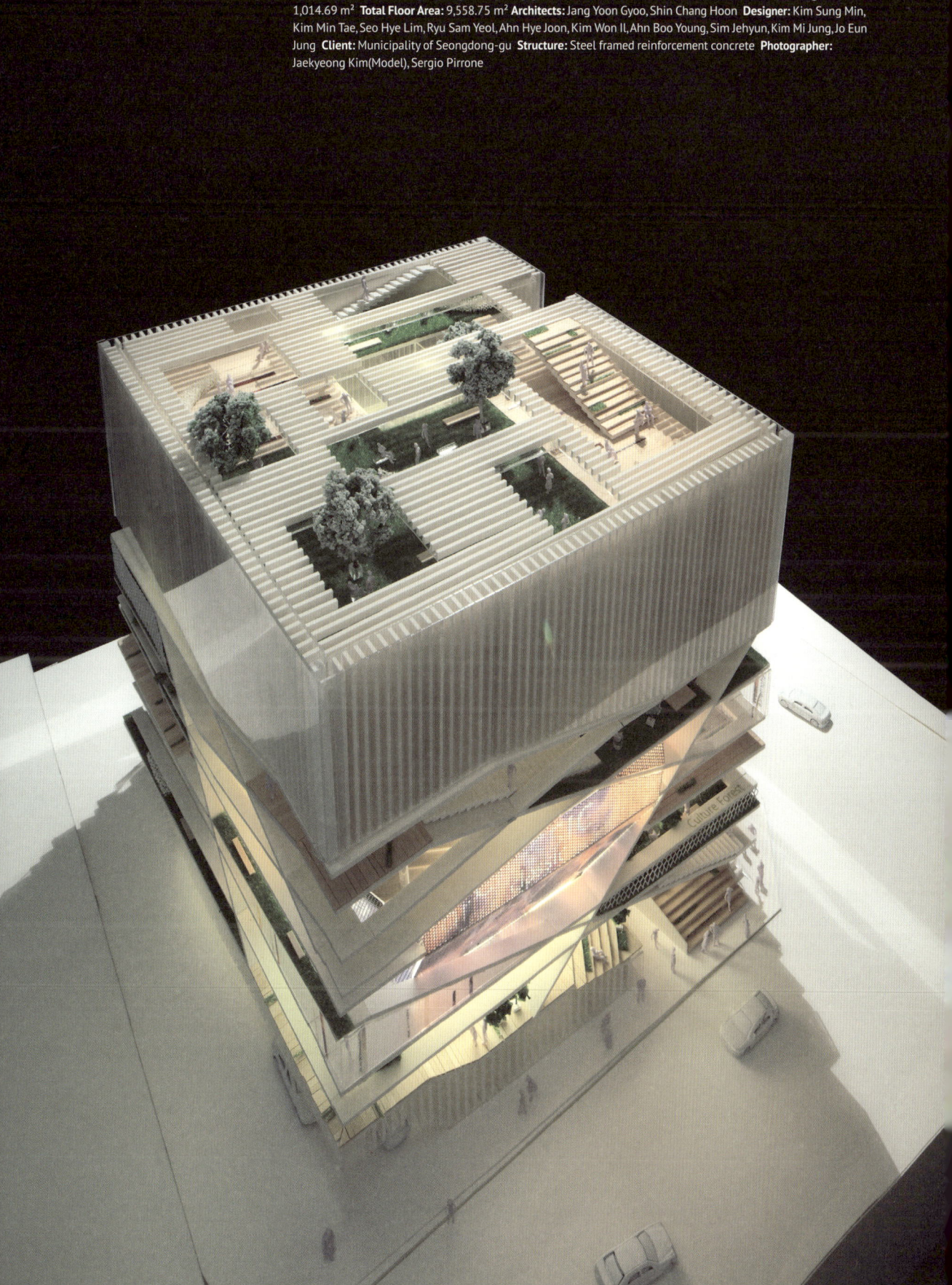

성동문화 복지센터는 단순히 관공서를 만든다는 개념이 아니고 열악한 도시환경에 새롭게 등장하는 문화복지시설의 역할을 담당하는 거버넌스의 개념을 수행한다는 점이 중요한 이슈가 된다. 성수동 공장지대에 가장 열악한 중심에 복지센터를 구성하고, 이 문화복지센터는 도시를 바꾸는 헤드쿼터와 같은 역할을 수행한다. 관공서의 기능이 행정위주의 역할이라면 거버넌스의 개념의 건축은 도시와 사회를 바꾸는 역할자로서의 씨앗과 같다. 성동문화복지센터도 공장지대의 주민들의 삶을 도와주고 재정비해주는 역할을 수행한다. 건축을 통해서 도시와 사회를 바꿀수 잇다는 전제에서 출발한다는 점이 흥미로우며, 빌바오 구겐하임, 그라츠 쿤스트하우스, 런던의 테이트모던, 라반센터, 알솝의 그리니치지역 도서관등이 그 예라 할 수 있다. 빈민지역에 씨앗처럼 뿌려져 주변사회를 바꾸는 성공적인 건축장치로 작용하였다.

Seongdong Cultural & Welfare Center is a crucial issue because it is not a simple concept of creating government office. It provides the governance concept of cultural & welfare center. The welfare center is provided in the center of the poorest factory district in Seongsudong, and this cultural & welfare center functions as the headquarters which changes the city. If the function of government office is mainly administration, the architecture of governance concept is like the seed of role which changes the city and society. Seongdong Cultural & Welfare Center also performs as the role of rearrangement and helps the inhabitants' life in the factory district. It was interesting premise that the city and society can be changed through architecture. And Guggenheim Bilbao Museum, Kunsthaus Graz, Tate Modern in London, Laban Centre, and Will Alsop's Peckham Library are the examples. It has been strewn like the seeds in the slum area and functions as a successful architectural device which changes the neighboring society.

The Evolution of Experiential Skin Building
경험적 스킨 구축의 진화

Jungwon Yoon (Department of Architecture, University of Seoul)
윤정원 (서울시립대학교 건축학부)

지금으로부터 15여 년 전, 2000년대 초는 모더니즘이 지배적이었던 90년대와는 조금 다른 시도가 건축잡지에 소개되기 시작했던 시기였다. 운생동의 건축을 처음 접했던 것도 그 때, 실험적 형태와 시스템을 제시하던 신생건축 그룹으로서였다.

    신체의 움직임으로부터 파생되는 단면들을 표면과 덩어리로 변환하여 연속적으로 이어나간 '바디스케이프'와 같은 설치작품도 있었는데, 운생동은 건물과 더불어 이러한 개념적 건축을 실험하고 실현하는 다양한 시도들에 대한 이론적 방향성을 정리한 『복합체』를 출판하기도 했다. 당시, 이종의 프로그램을 담는 공간을 구조 시스템과 복합적으로 통합해내는 방식이 개인적 관심사이기도 했는데, 이 책의 내용과 운생동의 건축적 선언은 그러한 관심과 딱 들어맞는 것이기도 했다.

    그 후 10여년이 훌쩍 지나간 지금 운생동은 한국 건축을 이끄는 대표 건축가 중 하나로 그들만의 독특한 형태적 언어와 접근법이 명확한 건축물들을 생산해내고 있다. 건축전공 학생들이라면, '성수문화복지센터'와 같은 건물들을 보았을 때 그 건축가를 지체 없이 '운생동'이라 지목할 정도로, 운생동 건축의 정체성은 확실하게 각인되어 있는 듯하다. 그것은 프로그램의 복합에서 기인한 단면의 형상과 그 형태적 특성을 외부로 노출시키는 방식과 관련된 것이라 생각된다.

A little more than 15 years ago, the early 2000s was a time when attempts slightly different from those in the 90s when modernism prevailed, began to be introduced in architecture magazines. It was at that time that I first encountered the architecture of UnSangDong, as a new architectural group that presented experimental forms and systems.

    There was also installation work such as 'bodyscape', which consisted of cross-sections derived from body movements transformed into surfaces and lumps and continued sequentially. In addition to buildings, UnSangDong has published "Compound Body", which summarizes the theoretical direction of various attempts to experiment and realize such conceptual architecture.[1] At the time, the way of integrating a space containing heterogeneous programs with the structural system was a personal interest of mine, and the content of this book and the architectural declaration of UnSangDong was something that fit perfectly with that interest.

    Ten years have passed since then, and today UnSangDong is one of the leading architects of Korean architecture and has been producing its own distinctive forms and architectures with a clear approach. It seems that the identity of UnSangDong architecture is imprinted surely enough so that when architectural students see buildings

그렇다고 해서 운생동 건축의 많은 부분을 앞서 기술한 특징만으로 설명할 수 있는 것은 아니다. 그간 사람들에게 많이 알려진 건축물만으로 운생동을 이해하려 한다면, 운생동이 지닌 건축적 역량과 가능성의 많은 부분들을 놓치게 될 수도 있다. 운생동의 여러 작품들을 고르게 놓고 볼 때, 2005년 『복합체』에서 제시되었던 각종 개념들은 지속적으로 다른 재료, 다른 시스템, 다른 프로그램, 다른 대지, 다른 장치 등을 통해 실험되고, 현실적인 부분들과 조율하면서 건축물로 탄생되어 가고 있는데, 이렇게 꾸준하게 스스로의 정체성을 만들어가는 일은 우리나라 현실에서 매우 어렵기 때문이다.

『복합체』에 제시되었던 주요 개념 중 '스킨(Skin)'이 있다. 스킨에 대한 아이디어는 내부로부터의 스킨, 프로그램스킨, 내장 없는 스킨, 장치로서의 스킨, 움직이는 스킨, 반응체로서의 스킨, 사이버 스킨, 미디어스킨, 지속가능한 스킨, 투명성의 스킨, 대지의 스킨 등으로 다양하게 제시된다. 이러한 스킨은 운생동이 도시, 건축, 공간, 프로그램 간의 복합체와 통합체를 구성하는 데에 매개적 역할을 담당하며, 복합체의 일부가 된다. 따라서 운생동의 스킨이 어떠한 "재료"를 사용하여 복합적 층위를 어떻게 구현하고 어떠한 구조, 공간, 프로그램과 연계되며 그것들이 건물에 대한 경험이나 도시에 대한 표정을 어떻게 변화시켰는가를 살펴본다면, 운생동 건축에 대한 담론에서 그다지 제기되지 않았던 재료와 구축이라는 관점을 더해나갈 수 있다.

운생동의 초기 설치, 전시작품을 보면 운생동은 작품의 재료를 선정하여, 그 특정 재료를 통해 형태와 공간을 입히기 위해 직접 시공, 설치 과정에도 몸소 참여함으로서 모니터 속이나 2차원 도면에 존재하던 복잡한 형태가 3차원 현실 속에 재현될 수 있도록 일조한다. 전형적인 재료나 시공법, 표준화된 상세를 사용하지 않은 경우, 디자인의 의도와 이를 구현하기 위한 설치방식을 완전히 파악하지 않은 경우, 일반 공사업체에서 설계안을 완벽히 구현해내기란 쉬운 일이 아니다. 그렇기에 건축가들이 작은 규모의 설치 작업을 진행할 때에는 좀 더 현장에 밀착함으로써 시공 과정에 적극 참여하여, 재료와 구축에 대한 경험과 지식을 축적하고 이를 건축 스케일로 확장시킬 수 있는 계기를 마련하고자 한다. 운생동은 서울시립미술관의 <물 위를 걷는 사람들> 기획전, <윤석화의 공간>에서는 천이라는 매우 가볍고 얇은, 형태가 불분명한 재료를 도입하여 스킨을 구성하고 건축적 공간을 생성해내는 시도를 했다. 건축가 스스로가 공사자로 참여하여, 몸소 천들의 고정 위치를 잡고, 재료의 케이블링을 통해 천을 잡아당기고 늘어뜨리는 방향과 길이 등을 하나하나 통제하며 설치해나갔을 것이다. 그렇지 않고는 천과 같은 재료를 정형화된 프레임이나 케이블 없이 통제하여 공간의 스킨으로 구축해내기란 쉬운 일이 아니다. 구현 재료와 시공 방식은 촉각적이다. 그리고 그 결과물은 하얀 천으로 이루어진 공간 속에 떠있는 구름과 같이

such as 'Seongsu Cultural Welfare Center', they are able to point out without second thought that the architects are 'UnSangDong'. I believe it is related to the shape of the cross-section resulting from the complexity of the program and the way of exposing its morphological properties on the exterior.

This does not mean that a large part of UnSangDong architecture can be explained only by the features described above. If you try to understand UnSangDong with only the well-known buildings, you may miss many of the architectural capabilities and possibilities that it has. When you look at various works of UnSangDong evenly, various concepts presented in Compound Body (2005) are constantly being tested through different materials, different systems, different programs, different sites and different devices, and in arbitration with reality it is being born as a building. This is because it is very difficult to make such a steady identity in the reality of Korea.

One main concept presented in Compound Body was 'skin'. Various ideas for skins are presented, from skins from the inside, program skins, skins without interiors, skins as devices, moving skins, skins as reactants, cyber skins, media skins, sustainable skins, transparent skins and skins of the earth. These skins play a mediating role in UnSangDong forming compound bodies and integrated bodies between cities, architecture, space and programs, and become part of the compound body. Thus, if you look at the "material" used by UnSangDong's skins, how it constructs a composite layer, how it relates to any structure, space, program, and how it changes the experience of the building or the expression of the city, a perspective of materials and construction, which has not been raised previously in the discourse on UnSangDong architecture can be added.

In UnSangDong's initial installations and exhibitions, they select the material of the work and participate in the construction and installation process to apply shape and space through the specific material, so that the complex form that existed in the monitor or two-dimensional drawing can be reproduced in three-dimensional reality. Without the use of typical materials, construction methods and standardized details, it is not easy for a general contractor to fully implement the design unless they fully understand the intent of the design and the installation method to implement it. Therefore, when architects install a small scale project, they are actively involved in the construction process by participating more onsite, accumulating experience and knowledge on materials and construction, and creating an opportunity to expand it to the architectural scale. in the exhibition "People Walking on Water" at the Seoul Museum of Art, and Performing Space for Yoon Sukhwa, UnSangDong attempted to compose skins and create architectural space by introducing fabric, a very light and thin material without a specific shape. The architect himself will have participated as a constructor, personally fixed the position of the fabrics, and controlled and installed the direction and the length of each piece of fabric, pulling on it and letting it down

비정형적이고 유연하게 늘어져있는 설치물로, 관람객은 그 안에서 몽환적이고 초현실적인 감각을 경험했을 것이다.

  이후, 운생동은 2005년 완공된 예화랑을 통해 파격적이고 아방가르드한 건축의 추구에 대한 자리매김을 확실히 한다. 예화랑의 스킨은 자유롭게 날아가는 듯이 접힌 형상을 구성하는 외벽과 건물의 실내 프로그램을 감싸는 또 다른 외피로 구성된다. 그 두 가지 요소 사이의 계단과 발코니 배열은 동선 공간을 내부로부터 밖으로, 전면으로 끄집어낼 수 있다는 명쾌하면서도 영리한 발상의 전환에 기인한다.

  5개의 판으로 분절된 외벽은 접힌 위치와 각도가 모두 제각기 다르다. 이 외벽은 사진만으로 접하거나 무심코 지나치는 경우, 일반대중이나 건축가들에게 노출콘크리트로 지어진 구조체로 오해가 되기도 한다. 그러나 이 외벽의 마감재로 적용된 베이스 패널은 시멘트를 압출 가공 성형한 판재로, 쪼개어 분절하고 이를 배열하는 방식에 의해 외관의 인상을 달리할 수 있다. 예화랑에서 베이스 패널은 수직 길이를 최대로 거의 1개층 높이만큼 확보하고 수평 폭은 1m 남짓 모듈화하고 있다. 이러한 비례로 인해 수직적 상승감이 더욱 강조되며, 자칫 밋밋해 보일 수 있는 면에 이질적이지 않은 슬릿 패턴들을 삽입하여 입면의 단조로움을 조금 깨뜨리려는 듯 보인다.

  베이스 패널과, 베이스 패널을 지지하면서 외벽과 건물 사이의 계단이나 발코니 등을 지지하는 철골 구조는 베이스 패널 안에 감추어져 있고, 외벽과 건물 사이를 가로지르는 철재 보는 색상과 처리 방식에 따라 외피의 부속물처럼 보이기도 하고, 외피로부터 독립적인 별도의 구성체 혹은 삽입물로 보이기도 한다. 특히 발코니 부분에서 철골보가 건물의 외피에 접하는 부분에 가로지르는 철재 부재를 두어 정방형의 완결적인 판이 구성된 것처럼 읽히게 하고, 그 위의 바닥판을 반투명하게 처리한 데에서 건축가의 세밀한 재료와 구축방식의 선정을 살펴볼 수 있다. 외벽과 건물의 내부 프로그램을 감싸는 외피 사이의 계단과 발코니 배열은 수직적, 수평적 켜와 틈의 겹침과 틈에서 발생되는 여러 가지 역동적 공간 및 시선의 교차를 만들어낸다.

  예화랑의 스킨은 외벽, 외벽과 건물외피 사이의 요소들, 그리고 건물의 실질적 외피를 통해 세 가지 층위를 구성하고, 그 스킨은 노출콘크리트로 구성된 구조체로서 착시되기도 하고, 도시와 건축 사이의 경계를 제어한다. 그러나 전면 스킨은 가로수길에 면하는 것이 아니라 좁은 접근로에 면하고 있어, 실제 도시 보행자에게는 틈 공간의 경험이 인지되고 외벽의 형태나 인상이 전달되기는 어렵다. 그에 반해 예화랑 이후 운생동이 강남에 실현한 건축물들은 매우 정직한 그리드 도시 격자와 고층건물들의 전형적인 규칙적 언어 사이에 이종의 스킨들을 적용하여, 도시 경험의 지루함에 일탈을 제공한다.

through the cabling of the material. Otherwise, it is not easy to create the skin of a space by controlling materials such as fabric without a standard frame or cable. Implementation materials and construction methods are tactile. And the result was an irregular and flexible installation hanging like a cloud floating in a space made of white cloth, in which the viewer would have had a dreamy and surreal experience.

  Since then, UnSangDong secured its place in the pursuit of unconventional and avant-garde architecture through the completion of Gallery Yeh in 2005. The skins of the gallery are composed of an outer wall that looks like folded shapes flying freely away and another outer skin that surrounds the indoor program of the building. The arrangement of the staircase and the balconies between these two elements result from a clear and clever change in the way of thinking that the circulation space can be pulled out from the inside to the front.

  The outer walls separated into five plates each have different folded positions and angles. if viewed through a photograph or without thought, this exterior wall can be misunderstood by the general public or architects as a structure made of exposed concrete. However, the base panel used as the finishing material for the outer wall is a cement plate that was extruded and processed, which can be change the appearance of the building depending on the way it is divided and arranged. In Gallery Yeh, the vertical length of the modulized base panel was maximized to be almost one storey high with a horizontal width of almost 1 meters. These proportions emphasize the feeling of vertical uplift, and the monotony of the elevation is broken by inserting not-heterogeneou slit patterns in the sides that could seem flat.

  The base panel and the steel structure supporting the base panel and the stairs and balcony between the outer wall and the building are concealed in the base panel. The steel bar traversing the exterior wall and the building may be seen as an appendage to the enclosure, depending on the color and treatment, or as a separate construct or insert independent of the enclosure. In particular, in the balcony, a steel material is placed to cross the part where a steel girder meets the outer skin of the building, allowing it to be read as if it is composed of a complete square plate, and the bottom plate above it is processed to be translucent. Through this we can see how the architect carefully chose the materials and construction method. The staircase and balcony arrangement between the outer wall and the enclosure surrounding the inner program of the building creates an intersection of various dynamic spaces and perspectives that occur in the vertical and horizontal layers and overlapping and gap of gaps. The skins of the gallery compose three layers through the outer wall, the elements between the outer wall and the skin, and the actual skin of the building. The skins can seem like structures composed of exposed concrete and control the boundary between the city and the architecture. However, the skin of the facade does not face Garosu-gil, but a narrow access road, which makes pedestrians recognize the the interspace but difficult to experience the shape and impression of the outer wall.

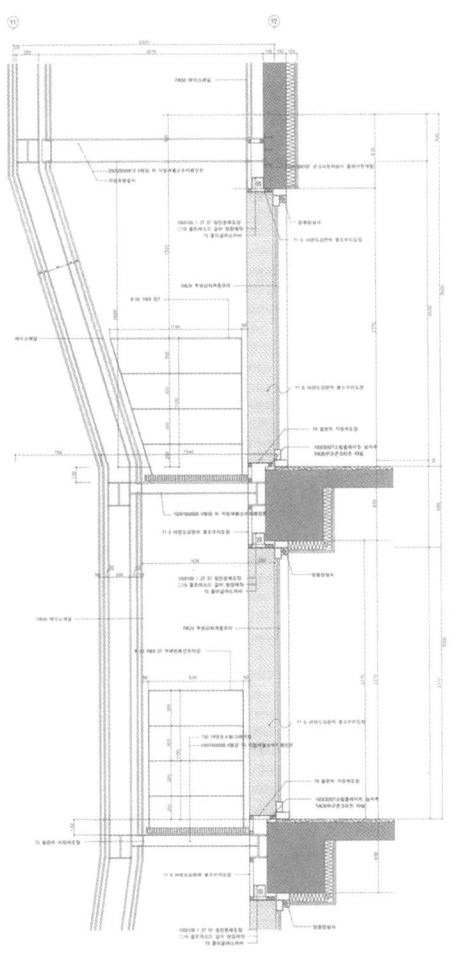

예화랑 단면상세도 (출처: 운생동건축사사무소)

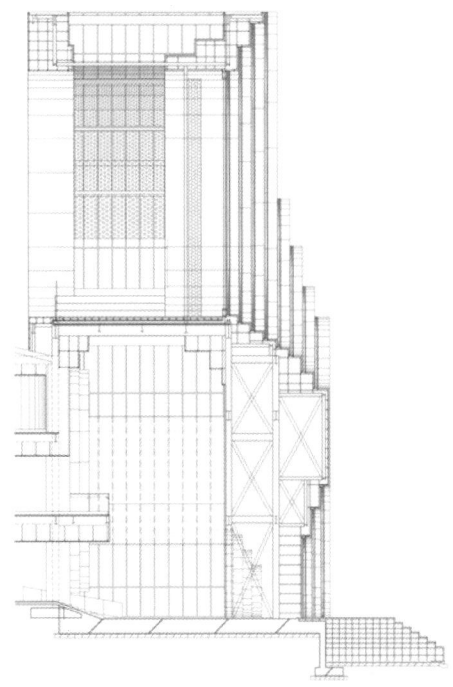

크링 단면상세도 (출처: 운생동건축사사무소)

2008년 운생동을 형태적 건축의 선두반열로 확실히 자리매김한 크링은 예화랑과 본질적으로 다른 형태를 지니지만, 스킨의 층위들 사이를 벌려 장치를 삽입하고 구조화시켰다는 점에 있어서는 비슷한 접근으로 해석할 수 있다. 크링은 현진 에버빌 사옥 계획안에서 제시되었던 브랜드 이미지를 표현하도록 제시된 외부스킨과 이를 구성하기 위한 복합체적인 프레임을 더욱 정교한 입체화와 패턴의 발전을 통해 현실화시킨 것이다.

크링의 스킨을 명확하게 구분하는 것은 불분명한데, 바로 이 지점이 건물을 흥미롭게 만든다. 전면의 높은 볼륨 자체를 두꺼운 스킨으로 보아야할지, 여러 동심원들이 내부로 파고들어가는 가장 외측의 조각적 표피를 이 건물의 스킨으로 볼지 판단이 어렵다. 이는 도시에서 관찰되는 건물의 표면은 동심원으로 구성된 외피로 인식이 되지만, 건물의 단면도를 살펴보면 외피와 내부 프로그램 공간, 그리고 후면의 낮은 매스들로 연결되는 부위의 외피가 일체화된 구조를 보이기 때문이다. 실제 스터디 모형과 공사 과정만 보더라도 이 볼륨은 동심원을 구성하기 위한

On the other hand, the buildings realized by UnSangDong in Gangnam since Gallery Yeh have applied heterogeneous skins between very honest grid city lattices and the typical language of high-rise buildings, thus providing a deviation from the boredom of urban experience.

Kring Kumho Culture Complex (Kring), which firmly established UnSangDong as a leading figure of formal architecture in 2008, has an essentially different form from Gallery Yeh, but can be interpreted as a similar approach in terms of how it opens the interspace between the layers of skin, inserting and structuring devices. Kring is a project that realizes, through elaborate three-dimensionalization and development of pattern, the exterior skins presented to express the brand image that was presented in the Hyunjin Evervill Office Building proposal and the composite frame to compose it.

Kring's skins are not clearly distinguished, which is what makes this building interesting. It is difficult to judge whether the high volume of the facade itself or the outermost sculptural surface where many concentric circles penetrate to the center should be regarded as the skin. This is because the outer skin composed of concentric circles is recognized as the the surface of the building observed from the city, but upon examining the section of the building, the outer skin, the inner program space, and the outer skin connected to the lower masses in the back are an integrated structure. Just the study model and construction process show that the volume is composed of a steel structure and

철골 구조체와 일체화된 구조로 이루어졌다. 가장 외측의 조각적 표피에도 내부로 삽입되어 들어가는 동심원의 깊이에 따라 두께가 부여되고, 그 안에는 승강기, 계단과 같은 동선 공간이 자리 잡고 있으며, 기능을 수행한다. 복합적인 스킨의 구축과 시공에는, 프리패브리케이션을 활용하여 스테인리스 스틸로 이루어진 가벼운 금속 외피가 도입되었다. 동심원으로부터 퍼져나가는 물결파장 모양의 패널 패턴과 오픈조인트로 패널 간 접합부위를 처리하여 패턴들을 한결 더 강조하여 보여준다. 또한 동심원을 채우는 재료들은 타공 금속판, ETFE 막, 커튼월 유리 등을 사용하여 각 원들이 서로 각기 다른 재현 방식의 미디어로 작동할 수 있도록 구현하였다. 그리고 동심원의 높이차에 따른 드러나는 측면도 ETFE 막으로 마감하여 야간에도 조명효과를 통해 동심원의 형태 언어가 강조되어 표현될 수 있도록 하였다.

반면, 최근 발표된 2010년대 운생동 건물의 스킨 구축은 그 방향성이 바뀐 듯하다. 스킨은 외부와 내부의 경계로서의 표피로만, 혹은 건물의 내부 프로그램과는 관계없는 도시에 대한 건물의 이미지로 국한되어 다루어진다. 장윤규는, "강남의 상업화된 중소 규모의 건축들은 건축이 가져야 하는 사회적 소통이나 공공성의 입장을 만들어내기 힘들며 표피적인 컨텍스트를 조장하는데 일조했다고 볼 수 있다"[2]고 기술한다. 스킨을 현실과 타협시킨 결과로서, 강남의 기존 그리드형 도시조직에 반하는 새로운 건축적 어휘를 표피에 도입한다. 그 예로, 몽유도원도 이상봉타워는 정방형 타워의 정면 위로 구형의 볼륨들을 배치하여 이형을 도입하고, 그 구들은 추상화를 통해 얇은 수직 루버들의 입면의 표피를 구성한다. 그리고 퓨처리즘그리드 미동전자는 리노베이션 프로젝트로서, 기존의 그리드 격자로 구성된 박스형 건물로부터 직교형 입면들을 거둬내고, 사선으로 구성된 입면을 덮는다. 그리고 기둥과 가장 바깥면 사이의 공간에서 획득할 수 있는 깊이를 통해 두께가 더해진 스킨을 구성한다. 지금은 컴퓨터를 사용하여 비정형적이고 불규칙한 패턴과 형태를 자유롭게 생산해낼 수 있으며, 이는 외피로 대체된다. 이러한 방식은 디지털 프로그램의 적용을 통해 건축설계를 하는 젊은 건축가들에게는 매우 익숙하고 대중적인 방식인데, 그러한 방식으로 생산되는 외피가 운생동 건축에서 비중이 커지며, 그들의 건축이 강남의 그리드형 도시 조직에서 이형으로 확연하게 존재감을 가지게 만드는 방식이 되었다.

이상봉타워와 미동전자 사옥에서 두 건물의 추상화되고 이질적인 표피를 구성하는데 쓰인 재료는 세라믹 박판이다. 세라믹이라는 재료는, 기존에 조적이나 타일과 같은 작은 단위체, 혹은 위생기기와 같은 볼륨 요소들로 쓰였던 반면, 근래에는 박판 기술이 발달하면서 그 크기가 다양해지고, 커지고, 하지 구조를 통해 매다는 방식을 통해 외피 재료로도 많이 쓰이고 있다. 이 두 건물이 세라믹 박판의 제품 광고에 대표적으로 등장하고 있는

an integrated structure to form a concentric circle. Thickness is given according to the depth of the concentric circles inserted into the inside of the outermost sculptural skin, and circulation space such as an elevator and a staircase is located therein. For the establishment and construction of complex skins, a light metal shell made of stainless steel was introduced using prefabrication. The joints between the panels are treated by ripple patterns spreading from concentric circles and open joints, resulting in an emphasis of the patterns. In addition, the materials filling the concentric circles were constructed using perforated metal plates, ETFE film, and curtain wall glass so that each circle could operate as media of different reproduction methods. The side, which was exposed due to difference in levels of the concentric circles, was also finished with the ETFE membrane so that the morphological language of the concentric circle would be emphasized at night through a lighting effect.

Meanwhile, the construction method of the skin in the recently announced UnSangDong's 2010 buildings, seems to have changed direction. Skins are limited as the boundary between the interior and exterior, or to the image of the building not related to the interior program of the building. Jang Yoon-Gyoo depicts, "It is difficult for commercialized small and medium-sized buildings in Gangnam to have the social communication or public position that architecture should have, and can be seen as contributing to the development of the exterior context." [2] As a result of compromising skins with reality, they introduce a new architectural vocabulary to the epidermis, which contradicts the existing grid-based urban organization of Gangnam. As an example, Mong-yoo-do-won-do Lee Sang-bong Tower introduces different forms by placing spherical volumes over the front of the square tower, and the spheres form the epidermis of the thin vertical louvers through abstraction. Also, as a renovation project, Futurism Grid Midong Electronics Headquarters is a renovation project that removes orthogonal elevations from a box-shaped building made up of existing grid lattices and covers the sloped facade. Then, the skin is composed with thickness added through the depth that can be obtained in the space between the column and the outermost surface. Now computers can be used to freely create atypical and irregular patterns and shapes, which is replaced by outer skins. This is very familiar and popular method among young architects who are engaged in architectural design through the application of digital programs. The outer skins produced in such a way take on more and more importance in the construction of UnSangDong architecture, and it has became a way for their architecture to make a distinctive presence with a heteromorphic form in the grid-shaped urban organization of Gangnam.

The material used to compose the abstracted and heterogeneous skin of both the Lee Sang-bong Tower and Midong Electronics Headquarters is ceramic sheets. Ceramic materials have traditionally been used as volume units, such as small units like masonry or tiles, or sanitary appliances.

것으로 보아, 새로운 재료의 등장과 이 재료의 정방형 모듈로부터의 탈피와 새로운 구축에 대한 욕구는 운생동의 건축을 통해 어느 정도 달성된 듯하다.

그러나 박판 재료를 외피를 감싸는 표피로 사용하는 방식이 예화랑이나 크링에서처럼 운생동이 스킨의 다중성을 실현할 수 있도록 입체화되고 구조화되고 이것이 재료의 선정이나 구축 방식의 발전으로 진화하지 않은 점은 사뭇 아쉽다. 이는 몽유도원도의 구형 볼륨을 구성하는 루버와 수직적 커튼월 사이의 접함과 사이 공간에 대한 질문이며, 미동전자사옥의 사선 그리드와 실내 오피스 공간 사이의 관계에 대한 질문이다. 기존의 재료를 새로운 부위에 응용하면서 맞닿게 되는 이질적 재료와의 접합이나 다른 역할로서의 구조적인 구축방식을 고민하고 발전시켜나가는 부분은, 우리 현실에서 녹록치 않지만 한편으로 도전해볼만한 목표가 될 수도 있다.

운생동이 1990년대 후반부터 한국 건축계의 아이디어 씽크탱크로서 지금의 3040 건축가들에게 많은 영감과 자극을 주고, 그들이 복잡한 형태에 노출되고, 거부감 없이 이를 체화하는 데에 기꺼이 한 획을 제공하였음에는 의문의 여지가 없다. 이제 또 새로운 도전의 단계로서 운생동이 좀 더 그들의 원색적인 개념과 형태를 재료와 구축의 진화적 선택과 적용을 통해 스킨을 구현하는 실행을 선도적으로 보여준다면, 실험적 구현의 선봉자로서 동시대 건축가들에게 지속적으로 자극이 되고, 건축을 지원하는 각종 기술 분야 관계자들에게도 새로운 방향을 함께 모색할 수 있는 협력자가 될 것이다.

Recently, with the development of thin sheet technology, the variety in size has increased; it has become larger and widely used as a covering material through the method of hanging it through the lower structure. As these two buildings are being typically used for commercials of ceramic sheets, the emergence of new materials and the desire to break away from square modules and to build new ones seems to have been attained to some extent through the architecture of UnSangDong.

However, it is quite unfortunate that the method of using thin sheet material as the skin covering the outer skin is not three-dimensionalized and structured for UnSangDong to realize the multiplicity of the skin as in Gallery Yeh or Kring, and that it has not evolved into the development of selection and construction methods of materials. This is a question about the tangency and interspace between the louvers and the vertical curtain walls that make up the spherical volume of the Mong-yoo-do-won-do. It is a question about the relationship between the oblique grid of Midong Electronics Headquarters and the indoor office space. The joining of heterogeneous materials that come into contact with existing materials as they are applied to a new part, or the part that we think about and develop a structural way of building as a different role may be a challenging goal in our reality, but worth giving a try.

UnSangDong, as a think tank for Korean architectural projects since the late 1990s, has provided a lot of inspiration and stimuli to today's 3040 architects, and there is no doubt that they have been exposed to complex forms and readily embodied them without a sense of resistance. Now, as a new stage of challenge, if UnSangDong leads the way in materializing skins through the evolutionary choice and application of materials and construction with more primitive concepts and forms, they will be a continuous stimulant to contemporary architects as the forerunner of experimental implementation, and will also be a cooperative partner, able to seek new directions for the various technological fields that support architecture.

---

1 장윤규, 복합체, 간향미디어랩, 2005, pp.16~17
2 장윤규, 논리적 오류의 풍경, SPACE 609 2018.08, p.39

1 Jang Yoon-Gyoo, Compound Body, Ganhyang Media Lab, 2005, pp. 16-17
2 Jang Yoon-Gyoo, A Scenery of Logical Errors, SPACE 609 2018.08, p.39

Hi Seoul Festival Stage Sculpture
Palace pf May_Thousand Palace

하아서울 페스티벌 무대 조각

Location: Seoul Plaza, Seoul, Republic of Korea **Allied Building around the site**: Seoul City Hall, Seoul Plaza Hotel, JEI **Architects**: UnSangDong Architects Cooperation **Principals**: Jang Yoon Gyoo, Shin Chang Hoon **Co-Artist**: Ahn Eun Mi **Structural Engineer**: MakMax Korea **Materials**: 3D MAK MESH (2 layered fabric made of translucent vinyl fiber) **Client**: Seoul Metropolitan Government, Seoul Foundation Arts and Culture **Photographer**: Sergio Pirrone

'5월의 궁'은 서울의 중심인 서울광장에서 '하이 서울 페스티벌 2009'의 '5월의 궁_천궁'은 거대한 환경 및 도시 조각일 뿐만 아니라 축제의 랜드마크이기도 하다. 이것은 60개의 직물로 구성되어 있다. 최대 길이는 약 200미터이다. 이것은 시청과 광장 주변의 건물들을 연결한다. 이 계획은 궁궐의 전통적인 장막 개념인 '용봉차일'에 착안해 만들어졌다. 이것은 모든 시민들이 왕과 여왕처럼 대접받아야 한다는 것을 의미한다. 이런 종류의 장막은 옛날에는 꽤 특별했다. 심지어 충신들도 이것을 사용하는 것이 허용되지 않았고 그것을 사용한 죄로 징역형을 선고 받았다. 축제를 위한 '5월의 궁'의 목적은 건물, 광장 및 거리를 연결하고 소통하는 것이다. 축제가 서울 광장과 함께 도시 전역으로 확대될 수 있게 해준다. 이 구조물은 하늘을 향해 치솟아 있다. 동시에 그것은 도시와 세계의 통합 지점을 만든다. 천연 재료를 적극 활용한 '5월의 궁'은 전통과 첨단 기술을 잘 조화시킨다. 바람은 건물에 나란히 불며 유동성에 기여한다. 흰색 반투명 천에서 들어오는 빛은 방문객들에게 멋진 공간을 선사하는 자연적 요소이다.

'Palace of May' comes into being a symbolic sculpture of 'Hi Seoul Festival 2009' in the heart of Seoul, Seoul Plaza. 'Palace of May_Cheon Goong' is not only a gigantic environmental and urban sculpture but also a landmark of this festival. It consists of 60 pieces of fabric. The maximum length of it is about 200-metre-long. They connects City hall, buildings around the plaza. The scheme is motivated by the concept of Korean traditional sunshade of the Palace, 'Yong Bong Cha Il'. It means that all the citizens have to be treated like the King and the Queen. This type of sunshade was quite special in bygone days. Even a loyal subject was not allowed to use this and was sentenced to jail due to asking for using it. The aims of the 'Palace of May' for the festival are connecting and communicating between buildings, square and streets. It enables the festival would extends all over the city as well as in the Seoul Plaza. This structure is soaring towards sky. Simultaneously it creates a convergent point of the city and of the world. 'Palace of May' mix well tradition and high technology. It made the best use of natural sources. Wind, which rises alongside buildings, that contributes to the fluidity. Light, which enters from the white translucent fabric, is the natural element that provides a spectacular space to visitors.

White Cube Matrix: Paju Kindergarten
파주출판도시 어린이집

Building Area: 495.62m² **Gross Floor Area:** 1,009.34m² **Building to Land Ratio:** 44.24% **Gross Floor Ratio** 90.10% **Building Scope:** 3F **Structure:** R.C. **Completion:** 2014. 7 **Architects:** Jang Yoon Gyoo + Seo Hyeon **Design Team:** Choi Young Chul, Kim Mi Jung, Koh Eun Jin, Yoon Ji Soo **Photographer:** Jaekyeong Kim(Model), Sergio Pirrone

유치원은 아이들이 꿈과 상상력을 키우는 공간이다. 아이들은 예측 불가능한 잠재력을 지니고 있다. 성장하는 세포들은 건축가 자신 스스로에게 고정된 형태와 공간을 정의 내리지 않고 미완성의 공간 확장을 만들어낸다. 다시 말해, 교실로 구성된 주요 단위인 흰색 큐브는 유치원의 기본 단위로 3차원적으로 쌓여 유치원을 이루고 있다. 이러한 큐브는 물질적 특징과 형태론적 완성이 제거되는 비물질적 공간의 속성을 만족시키는 것을 목표로 한다.

또한, 흰색 큐브 매트릭스는 인간, 자연, 문화, 예술, 물질 및 정신으로 구성된 추상성을 구성한다. 잠재력의 세포들은 시간의 흐름에 따라 그림과 낙서 같은 아이들의 상상력을 나타내는 다양한 매체로 채워질 것이다.

Kindergarten is a space that children create their dreams and imagination. Children bear unpredictable potential alike plain paper. Aggregation of growing cells generate uncompleted spatial expand without defining rigid form and space on architect's own initiative. In other words, white cube, the prime unit consisting of the classroom, are three-dimensionally stacked and complete whole body of kindergarten. These cubic cells aim to content the attribute of dematerialized space that material feature and morphological completion are eliminated. Also, the white cube matrix constitutes abstractness composed of human, nature, culture, art, substance and spirituality. The cells of potential would be filled with various media that represents children's imagination such as drawing and doodle with the elapse of time.

The Korean experimental architecture
한국적 실험 건축

Kyungsun Lee (Department of Architecture, Hongik University)
이경선 (홍익대학교 건축학과)

20년전 필자는 대학을 갓 졸업한 후 운생동의 장윤규 대표, 신창훈 대표와 같은 아르텍 건축설계회사에서 함께 일한 적이 있다. 당시에도 그들은 고뇌하는 젊은 건축가였으며 항상 일반인이 생각하지 못했던 새로운 컨셉을 제안했던 기억이 난다. 당시 나는 장윤규 소장이 소설가 이상과 비슷하다는 생각을 했었다. 이상이 기존의 문학적 체계를 무시하고 새롭고 실험적인 시도를 하였던 것처럼 그도 실험적인 건축 작업을 꾸준히 시도해오고 있다. 또한 그가 철학, 음악, 미학 등 건축뿐 만이 아니라 항상 다양하고 폭넓은 분야에 관심을 가지고 탐색하였던 것이 인상이 깊었다. 이런 백그라운드가 그들의 작업을 더욱 차별화시키는 밑바탕이 된 것이다. 필자는 하버드 박사과정 중이었던 2010년, 15년만에 미국 하버드건축대학에서 그들을 다시 만났다. 운생동의 강연을 통해 그들의 작품과 철학을 듣고 매우 감명을 받았으며 세계적인 건축가들이 초청받아 강연하는 곳에서 한국인으로서 특별강연에 서게 된 것도 자부심과 긍지를 느끼게 해주었다. 지금도 운생동의 작품을 보면서 필자는 20년전 느꼈던 그 실험적 접근 또한 결코 바래지 않고 지속되고 있음을 느끼게 된다.

Twenty years ago, I graduated from college and worked with architects, Yoon Gyoo Jang and Changhoon Shin in the Artek Engineering Architects Office. At that time, I remember that they were young architects who thought deeply and were always proposing new concepts that the general public did not think of. The director, Yoon Gyoo Jang, reminded me of Sang Lee, who is a novelist. As this author tried new and experimental ways to expand the boundaries of the existing literary system, Yoon Gyoo Jang has also been trying to work with experimental constructions. In addition, I was impressed that he not only explored philosophy, music, art, and architecture, but also explored a variety of other fields of interests. This background is the basis for further differentiating their works. It has been 15 years since I met them again at the College of Architecture at Harvard University in 2010 when I was a doctoral student. I was very impressed to hear about their work and philosophies through lectures by the architects of UnSangDong, and I was proud that they were Koreans and were invited to give a special lecture in a venue in which worldwide famous architects usually lecture. Even now, when I look at the works of UnSangDong, I feel that the experimental approach that I saw twenty years ago is also continuing without fading.

복합체 구축을 통해 한국다움을 보여주다

원래 한반도의 풍경은 결코 획일화되지 않고 단조로움을 배격하는 그 자체였다. 조상들의 건축물을 보면서 창호 조차도 그 어느 것 하나도 똑같지 않고 다름을 현대 문명사회의 효율 위주의 사상에서 보면 불량품이라고 할 수도 있겠으나 사실은 한반도의 풍경과 조화를 이루고자 했던 우리 조상의 지혜가 아닐까 싶다. 그런데 작금의 대한민국의 랜드스케이프는 그런 면에서 역사의 단절이라고 할 만큼 거기가 거기 같은 모습이다.

운생동의 건축은 복합체의 가능성을 실현하고자 하는 실험적인 건축이다. 그의 건축은 사람들이 체험하지 못한 새로운 복합체로 공간을 구현하고 도시에서 자리를 잡고 있다. 그런 의미에서 운생동의 작품은 부조화속의 조화라고나 할까? 어느 것 하나 비슷하지 않다. 그래서 재미가 있다. 오히려 마치 마른 땅에 피어나는 작은 떡잎처럼 획일화된 건물 숲에서 샘솟는 생수 같은 기분이 든다. 또한 획일화되기 쉽고 자칫하면 경제성의 논리로 타협하기 쉬운 현실에서 다른 곳에서 존재하지 않은 수공예적 건축작업을 탄생시키려는 건축가의 의지와 고집이 엿보인다.

복합체를 구축을 통해 공간과 장소를 만들다.

철학자 하이데거는 '장소'는 인간 실존이 외부와 맺는 유대를 드러내는 동시에 인간의 자유와 실재성의 깊이를 확인하는 방식으로서 인간을 위치시킨다고 하였다. 인문지리학자 렐프(Relph)는 장소는 개인과 공동체에 있어서 정체성의 중요한 원천이며, 때로는 사람들이 정서적·심리적으로 깊은 유대를 느끼는 인간 실존의 심오한 중심이 된다. 이러한 '장소'를 정의하기 위해서는 '공간'을 필요로 한다. 이프투안(Yi-Fu-Tuan)은 공간은 장소보다 추상적이며 무차별적인 공간에서 출발하여 우리가 공간을 더 잘 알게 되고 공간에 가치를 부여하게 됨에 따라 장소가 된다고 하였다. 즉, 공간을 이용하는 사람들이 그들의 경험과 기억, 기대, 꿈을 바탕으로 공간에 나름의 '의미'를 부여할 때 그곳은 장소가 된다. 공간은 사용자의 의미 있는 경험의 축적을 통해서 비로소 장소가 된다.

운생동의 작품은 복합체를 구축을 통해 공간과 장소를 만든다. 사람들의 기억에 남고 건축을 통해 장소를 만든다. 흔히 건축가들이 자신들의 디자인 언어를 반복적으로 구사하면서 매너리즘에 빠지게 되고 자신의 작품을 자신의 디자인 요소를 형태적으로 반복해서 사용하는 소위 셀프 레퍼런스(Self-Reference) 하는 경우가 많다. 그러나 공공건축물에서 사적 건축물, 대규모 프로젝트에서 소규모 프로젝트에까지 이르는 다양한 그들의 작품은 사람들의 기억속에 각인 되는 독특하고 혁신적인 조형미를 가지고 있다. 이는 프로젝트가 도시와 건축 그리고 사람을 담는 장소이어야 하기에 각 프로젝트마다 장소성을 형성을 위한 독특한 복합체의 형성물로

Compound Body shows what Korea is.

The scenery of the Korean Peninsula originally rejected a uniformity and monotonousness. If we look at buildings, we can say that none of the windows are the same, and they are defective based on the viewpoint of the efficiency of modern civilized society, but it might actually be the wisdom of our ancestors who wanted to harmonize with the landscape of the Korean Peninsula. However, the landscape of the present Korea is the same as that of the historical Korea.

The architecture of UnSangDong is an experimental construction that attempts to realize the possibility of a compound body. Its buildings create a space as a new compound body that people have not experienced and that take their places in the city. In this sense, the work of UnSangDong is a harmony in mismatch. None of the works are the same. So, it is fun. Rather, it feels like spring water in a uniformed building forest, like a small cotyledon blooming in a dry land. In addition, it seems to be the will and commitment of the architect who wants to create a handcrafted architectural work that does not exist elsewhere in reality that is easy to unify and easy to compromise with the logic of economy.

Compound body construction creates space and place.

The philosopher Heidegger stated that "place" positions human beings as a way of confirming the depth of human freedom and reality while revealing the bonds of human existence with the outside. Relph, a human geographer, said that "place" is an important source of identity in individuals and communities, and sometimes a profound center of human existence in which people feel deep emotional and psychological bonds. In order to define this "place," we need "space." Yi-Fu-Tuan said that space starts from an abstract and indiscriminate space rather than a place, and becomes a place as we become better acquainted with space and give value to the space. In other words, it becomes a place when people who use the space give their "meaning" to the space based on their experiences, memories, expectations, and dreams. Space becomes a place through the accumulation of the meaningful experiences of the users.

The work of UnSangDong creates space and place by building a compound body. It makes a place worth remembering through architecture. Architects often use their own design language and fall into mannerisms, and do so-called self-references, in which the user repeatedly uses his or her own design elements in a formally repeated manner. However, the work by UnSangDong, from public buildings, private architecture, to large and small projects, has a unique and innovative formative beauty that is engraved in people's memories. This appears as a unique compound body for a sense of place for each project, since the project should be a place that includes city, architecture, and people. In one of his major works, Kring, shocked me with an unconventional design which has not clearly been seen in other buildings. At the same time, the outer ring of the building is designed to wave to the inside and is

나타난다. 그의 주요 대표작 중 하나인 클링을 보면 분명히 다른 건물에서 보기 힘든 파격적인 디자인으로 충격을 주었다. 동시에 건물 외부의 울림통이 내부까지 파동이 이어지는 디자인으로 내외가 일관성 있게 구성이 되었다. 기업의 브랜딩인 '어울림'을 건축적으로 해석하면서 커다란 공명과 울림이 도시로 퍼트러 지는 것에 착안하여 기업의 정체성 확보뿐 만이 아니라 다각적으로 기업을 홍보하고 새로운 이미지의 창출을 꾀하였다. 또한 이러한 파사드는 외부와의 접촉을 통해 의사소통하고 역사 및 문화적 컨텍스트에서 정체성을 확립하였다. 이것이 내부의 프로그램까지 새로운 풍경을 통해 건축이 새롭기를 바라고 사용자들과의 소통, 문화와의 소통, 복합체로써의 장소를 만든다.

복합체를 구축을 통해 사회적 지속가능성을 보여주다.
한내 지혜의 숲은 도시재생이라는 측면에서 기념비적인 건물이라고 평가하고 싶다. 건물이 웅장하거나 특출난 디자인으로 사람들의 시선을 끄는 것은 아니지만 오히려 랜드마크성의 건물을 추구하지 않고, 작지만 목적성에 부합하고 이용하는 사람들의 니즈에 맞는 디자인이라는 점에 후한 점수를 주고 싶다. 이 작품은 내부에서 시작해서 외부로 전개해 나가는 정반대의 접근 방법을 취했다. 즉, 책장이 천장과 연결이 되고 다시 그러한 공간과 공간이 연결이 되어 내부에 다양한 높낮이의 차이로 다채로움을 선사한다. 도서관의 역할이 주이지만 동시에 풍부한 경험을 함께 할 수 있는 기회를 제공한다. 이는 대규모 랜드마크적 재개발 프로젝트를 통해 도시재생이 이루어진 것이 아니라, 작은 커뮤니티 공간의 변화를 통해 문화소외지역에 복합문화공간을 구축하고 이를 통해 도시재생을 이룬 것이다. 거대 도시 구조에서 볼 때, 작은 규모의 마이크로 스케일 건축을 통한 도시재생의 움직임 (Micro-scale Movement)의 파급 효과는 매우 크다. 그들이 보여준 장소성이 복합체구축을 통한 외부 파사드의 독창성, 내부공간의 디자인과 프로그램의 접합을 통해 나타났다면, 궁극적으로 사용자 간의 관계성을 통해 사회적 지속가능성을 보여주고 있다.

　　　　건물은 사람을 담는 그릇이다. 아울러 건물은 그 나라의 문화를 반영한다. 그래서우리는 세계 여러나라를 방문하면서 자연도 만끽하지만 아울러 그곳의 건물을 보고 감상한다. 외국 그릇에 담긴 구수한 청국장이 과연 우리가 기대한 그 맛을 낼 수 있을까? 신토불이 먹거리를 찾으면서도 외국 유명 건축가의 건물만 인정해 주는 모순적인 상황에서 운생동 대표는 신토불이 건축도 세계 시장에서 분명히 인정받을 수 있음을 증명하고 있다. 가장 우리다움이 가장 세계적인 것이라는 금언을 다시금 새기게 한다.

composed of consistent internal and external designs. By interpreting the company's branding, "harmony," and focusing on the spread of a huge resonance and echo to the city, the architects of UnSangDong not only secure a corporation's identity, but also promote the company in various ways and create new images. In addition, these facades also communicate through external contact and establish identities in the historical and cultural contexts. This point creates communication with users and culture, and a place as a compound body. I hope this will be architecture including a new inner program from a new landscape.

Construction of a compound body shows sustainability.
I would like to evaluate the Hannae Forest of Wisdom as a monumental building in terms of urban regeneration. I would like commend the fact that the building is not designed to attract people's attention due to its magnificence or exquisite design, but rather it is designed to meet the needs of people who use a building fitted to their purposes without seeking a sense of landmark for the building. This building took the opposite approach, deploying from the inside to the outside. In other words, the interior is filled with various differences of height as the bookcase is connected to the ceiling, and the space is connected to another space. This building plays the role of a library and provides an opportunity to have a rich experience. This is not an urban regeneration through a large-scale landmark redevelopment project, but is the urban regeneration through a change of a small community space, so it is constructed as a culture complex in a culturally marginalized area. From the viewpoint of a macro-urban structure, the ripple effect of an urban regeneration movement through the micro-scale movement is large. If the sense of place they showed is represented through a combination of creativity of the outer facade and an interior design via construction of a compound body, ultimately, the relationship between users shows social sustainability.

　　　　A building is a container for people. In addition, the building reflects the culture of the country. So, we visit many countries in the world and enjoy the nature, but also see and appreciate the buildings there. Can traditional Korean food in a foreign bowl really bring out the flavor we expected? In a contradictory situation where only the buildings built by famous foreign architects are recognized while people are looking for traditional Korean food, the representative architects of UnSangDong prove that traditional Korean architecture can be clearly recognized in the world market. This remark, that we are most global, is laid on my heart.

A Thousand City Plateaus
잠실 종합 운동장 재개발 국제교류복합지구

Location: Bogae-myeon, Anseong-si, Gyeonggi-do, Republic of Korea **Site area:** 8,204.9 m² **Building area:** 2,411m²
Gross floor area: 3,147.19m² **Building to land ratio:** 31.56% **Floor area ratio:** 36.51% **Building scope:** B1 ~ 2F
Height: 20m **Photographer:** Jaekyeong Kim(Model)

우리는 문화와 일상적 행사를 위한 새로운 컨테이너 역할을 하는 도시 고원을 구성할 것을 제안한다. 만약 도시의 새로운 도시 고원이 광장과 골목길들을 관통하며 도시 플랫폼을 만들고 있다면, 새로운 '도시 고원'을 만드는 것은 잠실에 새로운 프로그램, 새로운 풍경 그리고 주요 경기장을 관통하는 거대한 지도 작업을 하는 것과 같다. 서울 잠실 종합운동장 단지는 한강과 코엑스 사이의 단절로 인해 고립된 섬이 됐다. 우리의 제안은 코엑스 삼성역, 탄천, 한강 수상공원, 잠실 종합운동장을 강력하게 연결하는 도시 토폴로지의 재건이다. 연속적인 수평 평원을 통해 스포츠, 문화, 컨벤션, 엔터테인먼트 등 다양한 프로그램의 복합 공간 연속체를 제안 한다. 삼성 코엑스 일대에서 한강변까지 다양한 프로그램을 아우르는 수평 매트를 대표할 예정이다.

We propose to constitute a urban plateaus that acts as a new container for the culture and everyday events. If new city plateaus for the city is creating urban platform out of penetrations pathways and plazas, making the new 'city plateaus' is as same as to conduct an enormous mapping operation with penetrating new programs, new landscape and main stadium in Jamsil.
The existing block of the Jamsil Sports Complex in Seoul has been become an isolated island because of disconnection between the Han River and the Coex area as well as its deterioration. Our proposal is for the reconstruction of urban topology strongly connecting the Coex Samsung station area, the Tan Stream, the Han River waterfront park and the Jamsil Sports Complex. Through continuous one horizontal plateaus, we propose the continuum of compound spaces with diverse programs including sports, culture, convention and entertainment. It will represent the horizontal mat with diverse programs covering from the Samsung Coex area to the Han River waterfront.

Yeosu Expo 2012

여수 엑스포 2012

Architects: Jang Yoon Gyoo, Shin Chang Hoon, Kim woo Young, Kim Bong Kyun, Kang Seong Hyun **Photographer:** Jaekyeong Kim(Model)

자연상상체(Green-Imagination)는 자연(Green)과 상상력(Imagination)을 결합한 의미로 자연에 대한 무한한 가능성과 상상력을 현실로 변화시키는 힘을 재현하는데 있다. 여수박람회의 주제를 상징적으로 집약하여 전시하기 위하여 장소, 건축, 전시가 하나로 통합된 공간을 제시한다. 여수엑스포의 주제인 해양생태와 해양, 건축, 토목, IT 등의 첨단기술이 결합된 '해양 에코 복합체'를 구성했다. 바다와 연관된 천연과 인공적 자연의 경험과 전시를 담아내는 주제관을 구성한다. 전시관 자체를 자연을 형상화하여 자연의 일부분을 건축으로 가져오는 인공적 자연의 그릇을 제안했다.

'살아있는 바다, 숨쉬는 연안'이라는 주제를 대표하는 상징적인 건축을 구현하기 위하여, 바다의 정보적인 허브 역할을 수행하는 거대한 '바다의 문(Big Ocean Gate)'을 구성했다. Ocean Gate는 바다의 문화와 삶을 담고 있는 연안의 지형과 랜드스캐이프를 수직적인 모뉴먼트로 변환시키는 개념을 구성했다. '바다와 생명'이라는 주제를 상징적으로 보여주는 해양 에코와 건축을 결합한 수직적인 게이트가 된다. 수직적 랜드마크는 다양한 해양생태 공간적 경험을 구성하는 에코보이드를 구성하여 내부전시와 연계된 입체적 주제관의 역할을 수행한다. 여수박람회의 비젼과 메시지를 지속적으로 전달할 수 있는 모뉴먼트성을 구성했다.

Our emphasis is that by combining Green(Nature) and Imagination, we would make the best use of infinite possibility of nature. Having decided to show the exhibition space in a subtle way, we took rather unique way. We link the place, architecture and exhibition. We consider 'Ocean Eco Compounds' to realize the main theme of this exposition which consists of marine technology, architecture engineering, civil engineering, information technology and so on. We allow visitors experience ecosystem of the ocean and every kind of experience or exhibit which is related with the man-made nature in the Main exhibition area.

We set a place for festival of human being, ocean and technology. We set up a path for pedestrian which is extension of breakwater. It will take a role as a sort of a gallery which shows ocean exhibits in the Ocean exposition(Big O). Nonetheless, water was originally used as a horizontal component, Ocean Gate is cutting edge architecture which is vertical transformation of water. 'Big O', as core attractions for visitors, is connected with other exhibition facilities. There will be 'Eco-void space' on the extension of main hallway of exhibition area. In accordance with the changes and event of the eco-void space, visitors would be able to understand the main theme of this exposition.

Transformation of Compound Body to Social Imaginative Body

복합체에서 사회적 상상체로의 전이

Youngbum Reigh (Department of Architecture, Kyonggi University)
이영범 (경기대학교 건축학과)

더 윤리적인 건축을 향하여

인간의 삶을 풍요롭게 하는 가치를 파는 행위로서의 가능성을 건축이 내포하고 있다면, 건축에게 되묻는 중요한 질문 중 하나는 '공간의 윤리성을 어떻게 확보할 것인가?'일 것이다. 적어도 최소한의 윤리성을 획득하기 위해서 건축이란 문화적 헤게모니는 사회적 담론을 공간으로 재현하는 건축화의 과정과 태도에 늘 주목한다. 21세기 출발선이었던 2000년에 개최된 베니스 건축 비엔날레는 상당히 흥미로운 주제를 통해 건축화에 대해 전 지구적인 성찰을 촉구했다. 이 때 내 걸었던 '덜 미학적인, 그래서 더 윤리적인'(The City ; Less Aesthetics, More Ethics)' 이란 주제가 내포하는 의미는 무엇일까? 여기에는 도덕적 해이로 무장한 작가성, 그래서 자본과도 쉽게 결탁하는 작가성의 상업주의, 그리고 자신의 주관적 가치에 뿌리내린 문화적 코드를 사회에 강요하는 오만한 엘리트주의에 대한 심각한 경고가 담겨 있다. 하지만 건축은 자본과 사회, 개념과 구축, 공간과 행위라는 대립구도에 일정 정도 갇혀 있기에 미학과 윤리의 경계에서 외줄타기를 해야만 하는 태생적 한계를 갖는다.

도시에서는 존재 그 자체가 의미를 갖는 것이 아니라 자본의 흐름에 의해 존재가 끊임없이 순환되는 것이 절대 가치로서

Towards more ethics of architecture

The act of dealing with values which enrich human life, if this possibility is implicit in architecture, then, we must ask in return, 'how do you obtain ethics of space?'. In order to obtain minimum amount of ethics, architecture as regarded as cultural hegemony, often pay attention to architectural process and attitude which reproduces spatial discourse into space. Venezia Biennale 2000 at the starting line of the 21st century, called for self-reflection upon architecturalization in global scale by quite interesting subject. The subject 'The City ; Less Aesthetics, More Ethics', what is the implied significance? The subject calls for a serious warning against moral laxity in authorship which easily compromises with capital and against arrogant elitism which imposes subjective cultural code on the society. However, as architecture is confined in the opposite composition of capital and society, concept and construction, space and behavior, architecture bears limitation from its origin due to its position between ethical and aesthetical boundaries.

In the urban context, continuous cycling of the existence caused by capital flows dominate as an absolute value instead of valuing the existence in itself. Since nothing is stable in the urban context, therefore, it is absurd to ask for the root of spatial values in architectural

군림한다. 그래서 어느 것 하나 머물러 있을 수 없는 도시에서 건축이란 언어를 통해 공간가치의 근본을 묻는 것은 어리석은 일처럼 보인다. 하지만 변화하는 도시, 영속하는 것이 없는 사회에서 건축의 존재가치는 결국 변하는 것에 대해 어떻게 대응할 것인지, 그리고 그 변화의 힘과 어떻게 소통할 것인지에 대한 공간적 태도에 달려있다고 할 수 있다. 사회성의 작가적 해석에 근거한 건축화를 통해 미학과 윤리의 통합은 과연 가능할까? 이 질문에서부터 최근 운생동의 일련의 작업을 살펴보자.

복합체

복합체(Compound Body)는 지난 10여년간 운생동 건축작업의 토대를 이룬 개념이다. 건축적 개념과 공간적 실험으로서의 복합체는 다분히 작가성에 가까운 키워드이다. 꿈꾸는 가상의 세계를 실현하는 데 직면할 수 있는 위험, 그리고 그 위험을 끌어안고자 하는 욕망을 분출하는 작업으로서의 복합체라는 개념설정은 운생동 건축작업에서 무척 유효했다. 건축과 도시의 경계, 현실과 가상의 경계, 공간과 인간 행위의 경계를 샅샅이 탐색하고자 한 운생동의 그간 작업의 의미는 복합체의 건축화를 통해 결국 다양한 경계의 넘나듦을 경험할 수 있는 공간을 세상에 제안하는 데 있었다. 하나의 텍스트를 거부하고 공간적 변형과 조작을 통해 다양한 프로그램의 공간적 중첩을 만들어냄으로써, 하나의 텍스트로만 읽히는 공간의 스킨과 오브제를 중첩된 하이퍼텍스트적 공간으로 치환해내는 데 주력하였다. 물리적 실존을 뛰어 넘는 상상과 인지의 공간체로서의 복합체는 경험을 통해 인지의 공간이 갖는 한계를 파괴하는 하이퍼텍스트의 연속성으로 해석될 수 있다. 그렇다면, 운생동은 왜 이런 작업에 몰두하였을까? 복합체의 개념설정은 사회로부터 눌려져 있는 왜곡된 공간코드를 드러내서 사회화하는 작업으로 이해할 때 타당성을 갖는다. 공간을 대상화하지 않고 체험되는 객체로 설계하고, 사용자를 바라보는 객체가 아니라 체험하는 주체로 설정함으로써 공간과 인간의 복합체로서의 건축이 가능해진다. 또한 비현실의 세계를 공간을 통해 현실 가능한 세계로서 체험하게 하는 비현실의 재현이 복합체를 통해 이루어진다. 이 과정에서 개입되는 두 가지 중요하게 다뤄진 화두는 사회성과 상상력이다. 두 가지 화두를 바탕으로 끊임없이 공간을 체험하는 사람을 주체화해내고 공간에 개입된 건축가의 비현실적인 상상이 현실의 세계에서 창조적으로 재해석될 수 있는 여지를 갖게 만드는 이 불안정성이야말로 하이퍼텍스트로서의 복합체와 관계망이 만드는 경계면에 집중하는 운생동 건축작업의 창조적 동력일 것이다.

사회적 상상체

도시에서 존재하는 수많은 공간 관계, 즉 순응, 대립, 공존, 자생 등의 관계망의 다양성은 그 자체가 개념덩어리이다. 하지만 자세히

language. Nonetheless, architectural values within transforming cities of endlessly changing society, are dependant on spatial attitude towards ever-changing things and interacting with the forces behind the changes. I would like to introduce <UnSangDong Architects> projects by asking if it is possible to combine aesthetics and ethics through architecturalization based on authorial perspective on social nature.

Compound body

Compound Body is a concept which forms the basis of <UnSangDong Architects> projects for the past 10 years. Compound Body, practicing architectural concepts and spatial experiments, can be seen as a keyword close to authorial perspective. There is a danger in dealing with dreamlike fantasy world and the established concept, Compound Body, has been very effective in <UnSangDong Architects> projects as a way of releasing energy through work by embracing the danger. The significance of <UnSangDong Architects> projects lies in proposing space which allows frequent crossing between different boundaries through architecturization of Compound Body. <UnSangDong Architects> refuses single text and thus creates spatial layering of programs by transformation and manipulation. Their projects focus on substituting hypertext-like space by layering skins and objects for reading skins and objects as individual texts. Compound Body, imagination surpassing physical law of nature and recognizing spatial body, can be interpreted as continuity in hypertext which breaks the limits of recognized space from experience. So why did <UnSangDong Architects> devote to such work? The concept of the Compound Body has validity when understood as the process of socialization of initially discovering spatial code distorted by the society. Compound Body of space and human in architecture is made possible when first, space is designed as object to be experienced instead of objectifying it, second, when occupants are considered the main agent to experience. Also, regenerating unreality becomes possible through the Compound Body by providing experience of unreal world by space as real life experience. Here, social nature and imagination become two important topics in this process. There is instability here when architect's unrealistic imagination can be reinterpreted creatively in reality based on two topics mentioned above. And this motivates creative force in <UnSangDong Architects> works.

Social imaginative body

The diversity of relational network which exist in cities, numerous spacial relations such as adaptation, confliction, coexistence, self-generation, and etc., is itself a conceptual lump. But as we carefully look within, the diversity expressed in cities are only transformation and manipulation made by dichotomy of homogeneity and heterogeneity. The reason for feeling confliction and stifling in our stomach when we come across the urban boundary as we make relation with space, maybe lies here, eventually

들여다보면 도시가 연출하는 다양성은 결국 균질성과 이질성의 이분법이 만들어 내는 변형과 조작에 불과하다. 우리가 공간과의 관계를 통해 만나는 도시의 경계에서 갈등과 답답함을 느끼는 이유가 바로 여기에 있지만, 결국 이러한 대립적 구도는 공간의 새로운 형식의 제안을 가능하게 한다. 공간의 새로운 상상은 결국 특정하게 정의되지 않는 공간, 건축, 그리고 도시의 다양한 경계면의 가능성을 어떻게 연출해내느냐에 따라 성패가 좌우된다. 하지만 공간의 상상은 비단 건축가의 개념적 상상에만 의존하지 않는다. 공간에 개입된 힘, 즉 문화, 과학, 테크놀로지, 교통 등 이루 헤아릴 수 없는 물리적·기술적·사회적인 요소를 통해서 도시와 사회의 접점으로서의 건축을 찾아낼 수 있다. 한 걸음 더 나아간다면, 지속적인 역사와 문화를 바탕으로 테크놀로지(Technology)에 의해 필요에 따라 변화하는 임기응변적인 공간을 구성할 수 있다. 복합체의 연상선상이라 할 수 있다. '복합체의 궁극적인 목적은 무엇인가'를 다시 물어보자. 복합체가 불확정적이며 미완결체로서의 건축의 가능성을 끝없이 열어가는 하이퍼텍스트적인 공간이라면 결국 그 하이퍼텍스트의 경계면을 만드는 작업은, 공간의 본질에 대한 의문을 가속하는 전략으로서의 모호함과 공간과 사회의 명확하지 않은 접점의 가능성을 최대화하는 것이다. 본질이라고 믿는 것조차도 변화될 수 있는 불확실함이 공존하는 것이 지금 우리가 사는 도시(도시화된 사회)이다. 그 도시에 내재된 건축을 탐구하기 위해서 물리적 대상 자체에 관심을 갖기보다는 그들이 만들어내는 보이지 않는 관계에 주목하는 것이 지금 운생동 건축화의 기본태도이다. 대상 사이에 존재하는 그 무엇에서 도시와 건축의 본질을 발견할 수 있다고 보는 태도에서 우리는 사회적 상상체의 가능성을 엿본다. 사실 본질을 완벽하게 개념과 실체로 드러낸다는 것은 불가능한 작업이다. 하지만 이것을 가능할 수 있는 만드는 힘은 바로 상상력에 있다. 본질을 물성에 의존해서 드러내지 않고 또 개념을 작가성을 통해 구현하지도 않고 오히려 무한한 텍스트로 읽힐 수 있는 그런 관계망의 상상력을 공간화하는 작업이라면 오히려 본질 그 이상을 우리에게 보여줄 것이다.

사회적 상상체로서의 프로젝트

운생동 건축은 재생과 환경기계로서의 건축(regeneration & architecture as an environmental machine)에서 출발한다. 건축이 환경으로서 해석된다는 것은 결국 건축화된 공간이 갖는 가치의 순환체계에 기인한다. 공간 속에서의 관계의 순환체계를 만들어 낼 때 작가의 개념이 공간에서의 인간의 활동에 의해 재생된다. 이때 재생은 개념화된 공간의 본질이 인간 활동의 순간의 요구에 의해 관계망의 형태로 재현됨을 의미한다. 가치재생에서는, 건축에 의해 도달되는 미적인 조작보다는 새로운 도시환경의 변환을 유도하는 하나의 순환체계를 설정하는 것이

this conflicting composition makes possible a proposition for new spatial types. Fresh imagination regards to space, depending on how possibilities driven from various boundaries of unspecified spaces, architectures and cities, presents itself will determine the outcome. However, spatial imagination doesn't merely rely on architect's conceptual imagination. Architecture reveals itself as an interface between cities and societies through culture, science, technology, transportation, and many more physical/technical/social elements implied in space. Furthermore, it can create transmutative space by needs with technology based on continuous history and culture. It is in association with the Compound Body. 'What is the fundamental aim of the Compound Body?' We shall ask again. If Compound Body is indeterminate and hypertext-like space, then making the boundary of the hypertext maximizes ambiguousness as a strategy to accelerate questioning the essence of space and also maximizes possibilities made from uncertain interface between space and society. The society we live in is where indeterminate exist, even threatening our belief of what we believe as the essence. Instead of observing physical objects in order to study architecture of the city, <UnSangDong Architects> takes interest in invisible relations created by the city. We can watch for possibilities of Social Imaginative Body from their attitude. In fact, it is impossible to perfectly reveal the essence in concept and substance. But this can be possible by the power of imagination. Instead of revealing the essence based on material property, instead of regenerating concept with authorial perspective, but reading as limitless texts then space making of these relational network will provide more than the essence.

Projects as social imagination

<UnSangDong Architects> set off from regeneration and architecture as an environmental machine. To have architecture interpreted as in terms of environment arise from circulation system of values implied in architectural space. When creating circulation system of relations within space, architect's concept is regenerated by human activity in space. The regeneration mentioned here, identifies to conceptualized spatial essence being regenerated in the form of relational network through human activity based on instant demand. When regenerating values, it is important to propose a circulation system which induces change in new urban environment, rather than aesthetical manipulation reached by architectural intention. This circulation system allows endless regenerations of concept and significance, space and activity, relation and performance. When this circulation system is implied in space, architecture becomes a reactant. Being a reactant instead of a finished product, architecture creates many network between responsive objects. The network has made the boundary between architecture and society ambiguous. Architecture is not a terminated reaction of the society but it enables activities as an organic body.

중요하다. 이러한 순환체계가 결국 개념과 의미, 공간과 활동, 관계와 행위의 끝없는 재생을 가능하게 한다. 이 순환체계가 공간에 내재되어 있을 때 건축이 반응체가 된다. 완성체가 아니라 반응체이기에 결국 반응의 객체와의 사이에 수많은 관계망이 설정된다. 그 관계망에 의해 건축과 사회 사이의 경계가 모호하게 된다. 건축이 사회에 정지된 반응이 아니라 유기체로서의 움직임을 가능하게 하는 것이 바로 이런 이유 때문이다.

환경과의 소통체 & 가치재생의 사회적 상상체
텍스트의 공유, 가치의 재생, 영역의 통합, 공간의 혁신은 복합체에서 사회적 상상체로의 전이과정에서 보여준 운생동 건축의 일관된 태도이다. 여기서 보여준 프로젝트들은 건축 스스로의 생존뿐만 아니라 상호교류를 통한 공생관계를 설정하는 것이 지금 건축을 둘러싼 사회적 요구에 대한 반응이라 할 수 있다. 건축가의 개념에 의존한 피동적인 객체가 아니라 사회와의 관계망 속에서 무한한 정보를 수용하며 변형되는 자율적 주체로서의 건축이다. 건축을 통해서 행하여진 모든 문명의 흔적은 환경의 입장에서는 폐허를 의미하기도 한다. 미래적인 요구에 의해 문명이라 일컬어지는 것들조차도 이제는 죽어버린 건축적 장소와 공간을 만들어 나간다. 사회가 더욱 문명화 될수록 구축화된 모든 공간체제로 인해 자연환경은 폐허를 향해서 나아간다. 무엇이 본질이냐에 대한 의문을 가져야만 하는 시점에서 운생동의 화두는 복합체를 넘어서서 가치의 재생을 위한 사회적 상상체로 귀결된다. 사회가 요구하는 건축화의 본질이 무엇이냐를 묻는 작업이다. 이 작업에서 가장 먼저 요구되는 태도는, 건축을 경험하고 소비하는 주체와 건축이 공간을 통해 끊임없이 재생산하는 객체 사이에 존재하는 반응에 주목하는 일이다. 보이지 않는 수많은 관계들을 어떻게 드러내느냐, 그리고 본질에 가장 가까운 상황을 공간적으로 어떻게 연출하느냐에 따라 건축은 수많은 관계망에 의해 이루어진 상황(circumstances)으로 전환된다. 인간, 건축, 도시 등의 전체적 설정에서부터 이들을 구성하고 있는 각 요소들에 이르기까지 우리의 삶은 내·외적인 관계들에 의해 정의된 동적 구조(dynamic structure) 속에 존재한다. 결국 건축에게 주어진 과제는 이러한 관계를 상호 연결시켜주는 장치를 만들어내는 것일 뿐이다. 운생동의 건축적 태도는, 근본적으로 새로움을 통해서 건축과 그 외에 함께 관련된 영역의 정의와 통합을 새롭게 만들어낼 수 있다고 믿는 듯하다. 전통적인 공간의 개념이라든지, 공간을 규정하는 개념적인 부분에서 물리적인 요소까지 전반적인 부분에서의 의문으로 시작하여, 지금까지 존재하지 않았던 다양한 부분의 건축을 발견하기를 원하는 것이다. 여기 소개된 4개의 프로젝트를 포함하여 운생동의 연속적인 작업이 보여준 건축화의 일관된 태도는, 작가성이 갖는 신화적 상상과 테크놀로지와 결합된 사회적 상상을 통해 구현된 가치재생의 사회적 상상체를 드러낸다.

Environmental communication & social imagination of value regeneration
<UnSangDong Architects> shows us consistent attitude of text sharing, conceptual regeneration, integration of domain, innovative space, towards architecture by transformation of the Compound Body to Social Imaginative Body. From introduced projects above, not only the survival of architecture itself but also proposing co-existence through interaction are reaction to social demand which surround architecture today. Not an architect's role as passive object dependant on concept but an architect's role as autonomous subject where it transforms by receiving infinite information within social network, this is <UnSangDong Architects>. The evidence of civilization made through architecture signifies a ruin in environmental perspective. So called civilization demanded by futuristic needs, even these bring death in architectural site and space. Construction of all spatial system due to ever more civilized society, the natural environment is becoming a ruin. At this point the question of 'what is the essence?' is necessary, the topic of conversation at <UnSangDong Architects> goes beyond the Compound Body and into Social Imaginative Body in order to regenerate values. This questions the essence of architecturalization society demands. The first expression needs to pay attention to reactions existing between the main agent who experiences and consumes architecture and numerous objects architecture regenerates through space. How do you reveal endless invisible relationship network? How do you spatially perform the closest circumstance to the essence? Depending on the answers to these questions, architecture transforms into circumstances made of endless relationship network. From the whole set-up of human, architecture, city and to its elements composing it, our life exist within a dynamic structure defined by internal/external relations. Eventually the task given to architecture is merely creating a device interconnecting these relations. <UnSangDong Architects> architectural perspective believes that basically through newness, new creation is made possible by redefining and combining architecture and its related field. Overall their work begins in questioning, from questioning traditional spatial concept to newly discovering architecture un-identified in the past. <UnSangDong Architects> has shown consistent attitude of architecturalization through their continuous work throughout including 4 projects introduced above. It reveals Social Imaginative Body which regenerates values by combining mythological imagination and social imagination.

Shanghai Expo 2010
상하이 엑스포 2010

Location: Paldal-gu, Suwon-si, Gyeonggi-do, Republic of Korea **Site Area:** 7,504.30 m² **Building Area:** 3,945.66 m² **Total Floor Area:** 28,934.62 m² **Structure:** RC Structure **Architects:** Jang Yoon Gyoo, Shin Chang Hoon **Photographer:** Jaekyeong Kim(Model)

녹색 도시와 녹색생활이라는 주제와 일치하는 Korean Corporate Pavilion을 상상한다. 이는 Communi-Imagination이라고 불리며, 기술 혁신과 함께 환경에 대한 내성을 가지고 있다. 한국은 한국전쟁이라는 불행한 상황을 극복하고 상상할 수 없는 발전과 혁신을 이룩했다. 이 공간은 이러한 업적의 주역인 한국 기업의 기술과 정신을 상징한다. 정보기술, 유통, 항공, 전자, 자동차, 화학, 조선 등 첨단기술로 발전하고 있는 12개 한국 기업의 정신과 기술은 다양한 전시와 비디오로 건축물에 표현된다. 이 기록적인 규모의 엑스포는 세계에서 가장 큰 시장인 중국에서 개최된다. 우리는 선진적인 한국의 기술과 브랜드를 홍보하고, 한국과 중국의 경제와 문화 교류를 위한 새로운 관계를 재구축하기를 바란다. 자연과 한국의 첨단 기술에 대한 새로운 인식을 가지고 한국 기업들이 새로운 경쟁력을 증진시키기를 희망한다.

We imagine Korean Corporate Pavilion coinciding with the subjects; Green city and Green life. It is named Communi-Imagination and it holds introspection of environment together with innovation of technology. Korea got over the unfortunate situation of the Korean War which didn't seem to be possible and has achieved unimaginable development and innovation. This space represents technology and spirit of Korean enterprises which is the main agent of these accomplishments. The spirit and the technology of 12 Korean enterprises which are developing towards higher-tech such as information technology, distribution, aviation, electronic, vehicle, chemistry and shipbuilding are represented in the architecture by diverse exhibitions and videos. This recordable scale of expo is held in China which is the biggest market in the world. We hope to promote advanced Korean technology and brand, and re-establish a new relationship for economy and cultural exchange between Korea and China. We hope to promote new competitiveness of Korean enterprises with new awareness of nature and Korean high-technology.

Mogyudowondo(Imagination City)
몽유도원도(상상의 도시)

Architects : Jang Yoon Gyoo, Shin Chang Hoon + Mak Max Korea **Photographer:** Jaekyeong Kim(Model)

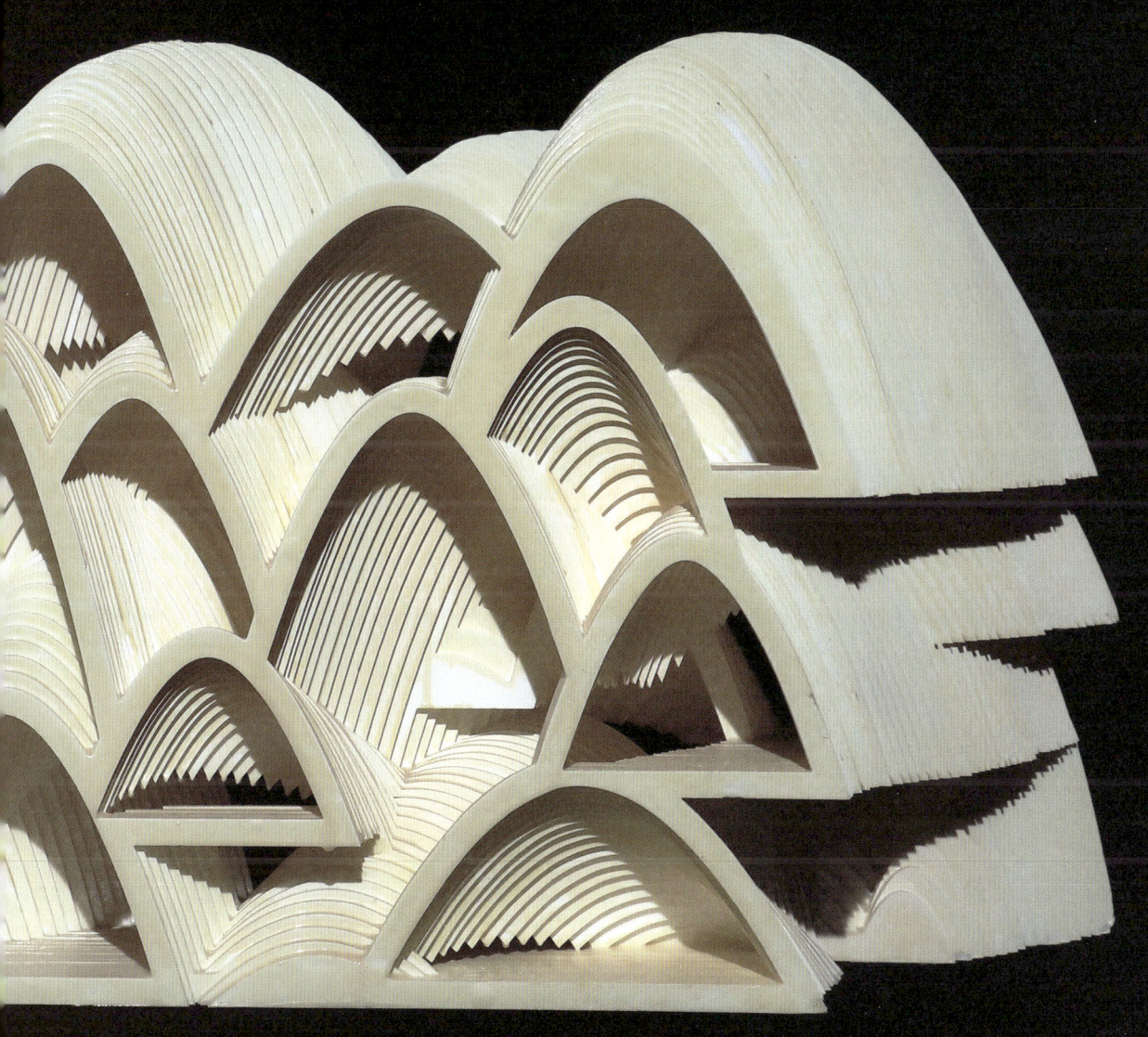

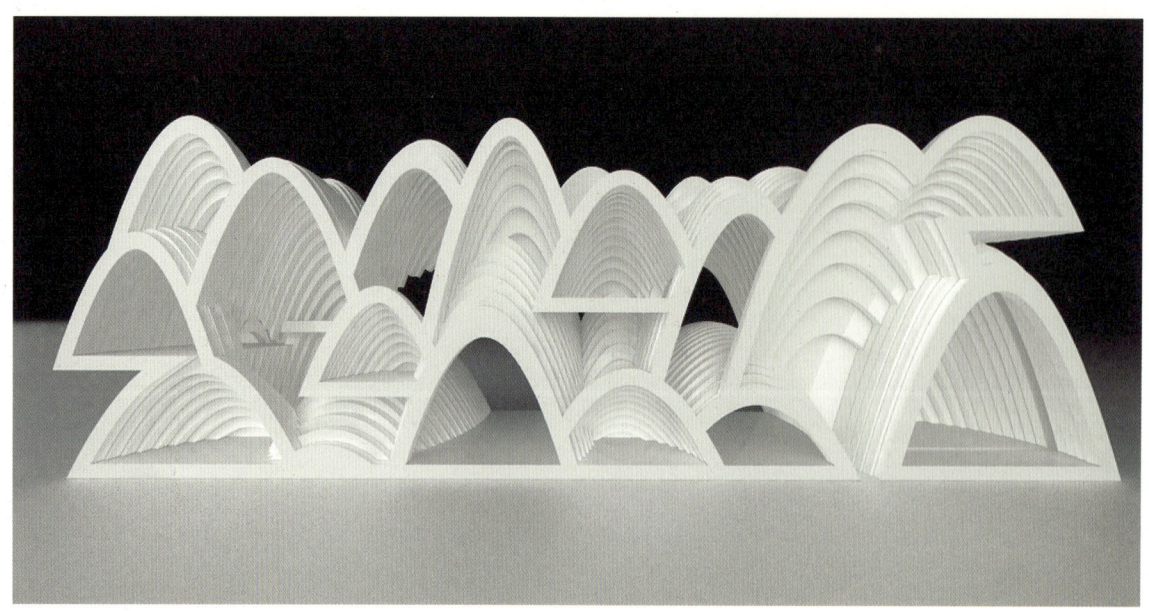

동양의 풍경화는 이상적인 풍경을 그리는 것이다. 이것을 들여다 보는 것은 흥미롭다. 서양 화가들은 자연을 그대로 묘사하는 반면 동양 화가들은 그들의 기억을 바탕으로 자연을 재현한다. 동양 예술가들은 자연과 그들의 상상력의 일부를 묘사한다. 그것은 자연의 지속가능성을 반영할 뿐만 아니라 새로운 유토피아를 만들어내려고 노력한다. 동양 예술가들은 인위적이지도 그렇다고 허구도 아닌 자연의 상상력을 지닌 새로운 근본적 건축을 꿈꾼다. 인위적인 구조의 자연은 영원성을 가졌으며, 한국 풍경화의 변화를 가져온다. 풍경화는 동양 자연주의의 관점에서 시간의 흐름과 과정을 가진 <시각적 매체>로 설명 할 수 있습니다. 저자는 자연과 끊임없이 변화하고 호흡하는 건축물이자 살아있는 창조물인 <인공의 도시 – 상상으로 보는 도시>를 제안한다. 인공 악기는 자연을 시각화하는 다양한 방법으로 만들어졌다. 지형의 3차원적 조합은 다양한 결과를 낳는다. 한국의 산, 들판, 계곡, 물과 같은 다양한 특징들을 추상화하여 건축과 결합한다.

Oriental landscape painting is about drawing an ideal scenery. It is interesting to see; the Western painters depict nature exactly as it is while the East ones reproduce nature based on their memory. The oriental artists depict nature with fraction of the memory and their own imagination. Not only reflect they sustainability of nature, but also try to construct a new utopia. Oriental artists dream new fundamental architecture with an imagination of nature which is neither artificial nor real nature. Nature in an artificial structure also has eternity and changes of Korean landscape painting. Landscape drawing can be explained as <visual media> that has trace and process of time in view of oriental naturalism. I suggest <Artificial nature-city as an imagination> which is an architecture, and also a living creature, that is constantly changing and breathing with a nature. The artificial instrument is constructed in a diverse ways of visualizing nature. The three dimensional combinations of topography result variety outcomes. I abstract diverse features such as Korean mountain, field, valley, water and combine them with architecture.

The Light and Shade of Skin Architecture
표피 건축의 명암

Eunseok Lee (Department of Architecture, Kyonggi University)
이은석 (경기대학교 건축학과)

운생동 건축에 대한 첫 기억은 지금으로부터 12년 전 예화랑의 인상깊은 파사드로 거슬러 올라간다. 예화랑 입면의 날카로우면서도 강력하게 도심을 뚫고 솟아오르는 역동성과 구성의 세련미는 국내외 저널리즘을 통해서 주변인들이 의식하기에 충분했다. 최근에는 거의 국외 건축물만을 다루고 있어 그 특성을 달리하고 있지만, 당시 국내 건축가들의 작품으로 매월 특집을 구성하던 건축저널 C3korea의 2006년 1월호 표지를 예화랑의 이미지가 뚜렷이 장식하면서 즉시 운생동이라는 생소하던 건축 아뜰리에의 이름을 국내외에 알렸다. 그렇게 건축그룹 운생동이 존재감을 드러낸 후로부터 12년, 한국산 국제 건축전문 저널인 Space의 2018년 8월호 표지에 또다시 운생동이 설계한 이상봉 타워의 크로즈업 된 사진이 게재 되었다. 비록 예화랑과 이상봉타워 양 건축의 컨텐츠가 서로 다르고, 개축과 신축이라는 시공의 환경에도 차이가 있으며, 준공 시점을 보아도 무려 십 수년 이상의 꽤 긴 시간적 간격을 지니고 있다. 그러나 공교롭게도 이 둘은 겉으로 보기에 매우 유사한 건축적 개념의 반복 재생이라는 인상을 떨칠 수 없다. 이를 두고 최근 한 비평가는 "편린이 반복되는 추상화된 표면"이라고 부르고 있다. 운생동의 건축은 과연 편린이 반복되는 것인가, 아니면 표면이 반복되는 것인가?

My first memory of UnSangDong Architects Cooperation dates back 12 years from now, to the impressive facade of Gallery Yeh. The dynamism and the sophistication of the composition of Gallery Yeh's facade, which shoots up swift and strong through the center of the city were enough to grab the attention of neighbors through domestic and foreign journalism. There was a period when architecture magazine C3 Korea, which in recent years features mostly international architects, published monthly special features of works by domestic architects. When an image of Gallery Yeh was printed on the cover of the January 2006 edition, UnSangDong, the name of an unknown architectural atelier, gained immediate fame at home and abroad. 12 years after the architectural group UnSangDong rose to the surface, a close-up image of their Sang-bong Tower was printed on the cover once again, this time on the August 2018 issue of Space, the Korean international architectural journal. Although Gallery Yeh and Sang-bong Tower differ from each other with different contents, difference in the construction environment (reconstruction and new construction), and there is a long interval of more than several decades between the time of completion, coincidentally, I cannot shake off the impression based on the exteriors that it is play on repeat of seemingly very similar architectural concepts. A critic recently called this "the abstracted surface

운생동의 예화랑은 지형건축 (Land Architecture)의 트랜디한 수평적 이미지를 마치 서울 강남의 도심에 수직으로 우뚝 세워 놓은 듯 했다. 사실 지난 세기 말에서부터 21세기 초까지, 지형건축의 경향성은 마치 쓰나미와 같이 세계 건축계에 유행처럼 번지게 되었다. OMA의 로테르담 쿤스트 할과 파리 쥬시유 대학도서관을 필두로 해서, MVRDV가 시도한 여러 건축물과 무명의 건축가 자에라 폴로를 세계가 주목하게 만든 요코하마 페리 스테이션 국제현상의 당선작과 같은 것이 대표적인 예가 되겠다. 한 때 한국에서도 그 영향은 뚜렷했는데, 헤이리에서 실험되었던 조민석의 '딸기가 좋아'나 김준성, 김종규 설계의 '헤이리 주민센터', 그리고 최근까지 시도되는 곽희수의 몇몇 주거 프로젝트에서도 그러한 경향성은 뚜렷이 두드러지곤 한다. 이처럼 지형건축은 대지의 형상을 인공으로 조성하며, 융통성 있는 프로그램과 더불어 자연 대지가 지니는 가변적인 성질을 건축 내외부로 끌어들여 융합함으로써, 역동적인 형태 및 공간효과를 극대화하려는 설계 전략이다. 수평과 수직으로 엄격히 분리되었던 고전적 공간의 경계를 허물며 인접 공간을 연결지어 흐르게 만들었던 모더니스트들의 욕망을 더욱 심화시키고, 공간의 수평적 상호관입 정도만을 시도한 자유평면 방식을 넘어, 수직의 층간 경계까지도 해체하고 뒤섞음으로써, 궁극적 건축공간의 자유에 도달하려는 일련의 의도이다. 이는 데카르트적 합리성과 몬드리안 구성의 한계에 갇혀있던 근대주의 공간의 개념을 흔들어 전복하려한 시도요, 플라토닉 모더니즘의 단순한 볼륨이 누리던 절대적 지위를 무너뜨리려는 참신한 돌파구이기도 했다. 이처럼 지형건축은 서로 다른 공간의 영역을 섞고 통합하여 사용자가 상호 소통하며 궁극에는 건축 프로그램을 도시 속에서 서로 공유해 보고자 하는 도전적 시도였다. 이는 하이브리드와 융합의 시대로 대변되는 21세기적 시대정신과도 자연스럽게 일치하는 가치이었다.

마치 중첩된 대지의 형상이 전방위로 변모하는 것처럼 예화랑에서의 수직 지형 변화는 서울 도심의 하늘을 파고들며 너울댔으며, 주변 도시 골목길은 계단실의 틈새를 비집고 들어와 소통하려 했다. 예화랑은 발코니마다 다양한 도심의 전경을 프레임 속으로 재단해 끌어들였고, 접히고 펼쳐지며 꺾이는 조형적 파사드의 놀이를 통해 동일한 재료를 사용했음에도 불구하고 내외부와 상하부로 넘나들며 변화무쌍한 효과를 보였다. 비로소 건축가의 의도대로 스킨이 바닥이 되고 바닥이 천정으로 변화하며 공간을 활성화하는 상태가 된 것이다. 그리고 그 역동적 볼륨은 주변 강남의 빌딩들과도 사이공간을 서로 견주며 공유하는 듯 보였다.

이와같이 태생적으로 예화랑과 매우 유사한 DNA를 지니고서 십 수년이 지난 이후에 가까운 강남에 태어난 이상봉타워 역시 기존 도시적 시선의 틀을 깨고서 애써 새롭게 건축적이고

of a recurring bit". Does the architecture of the UnSangDong repeat a small part of itself, or is it the surface that is repeated?

UnSangDong's Gallery Yeh seemed to make a trendy horizontal image of Land Architecture stand vertically in the city center of Gangnam, Seoul. In fact, from the end of the last century to the beginning of the 21st century, the tendency of land architecture has spread like a tsunami throughout the architecture community all over the world. Headed by OMA's Kunsthal in Rotterdam and Jussieu Library in Paris, some representative examples would be several works by MVRDV, and the winner of Yokohama International Passenger Terminal Design Competition, which attracted the world's attention to the unknown architect, Zaera-Polo. The impact was once evident in Korea as well. Such tendency is obvious in Minsuk Cho's 'I Like Dalki', which was an experiment in Heyri; 'Heyri Community Service Center' by Junsung Kim and Jong Kyu Kim; and some of Hee Soo Kwak's ongoing residential projects. In this way, land architecture is a design strategy to maximize the dynamic shape and spatial effect by artificially forming the shape of the earth, and by integrating the variable nature of earth along with flexible programs with the inside and outside of the building. It is a set of intentions to reach the ultimate freedom of architectural space, by breaking down the boundaries of the classical space that was strictly divided horizontally and vertically, and deepening the desire of modernists who connected adjacent spaces and made them flow; going beyond the free-plane method which attempts only horizontal interpenetration of space and disassembling and shuffling even the vertical interlayer boundaries. This was an attempt to shake and overthrow the concept of the modernist space, which was confined to the limitations of cartesian rationalism and Mondrian composition, and also a novel breakthrough attempting to defeat the absolute position enjoyed by the simple volume of Platonic Modernism. As such, land architecture was a good attempt to mix and integrate areas of different spaces for users to communicate with each other and ultimately share architectural programs with with one another in the city. It was a value that naturally coincided with the spirit of the 21st century, represented by the hybrid and fusion era.

Just as if the shape of the overlapped earth was transformed all over, the change of vertical topography in Gallery Yeh was making waves over the sky in the center of Seoul, and the surrounding urban alleys entered the gap in the staircase and tried to communicate. Gallery Yeh pulled in various city views into the frame of each balcony, and through playing with the facade by folding, unfolding and bending, it showed a variety of effects in the interior and exterior, upper and lower, despite using the same materials. At last, the skins become floors and the floors change into ceilings, activating the space according to the intention of the architect. And the dynamic volume seemed to compare against and share the interspace with the surrounding Gangnam buildings.

The Sang-bong Tower, which was born in nearby

도시적인 가치에 도달하고자 한다. 하지만 좁은 골목길을 비집고 틈새에 놓여야 했던 예화랑 시절보다도 이제 훨씬 더 큰 대로변으로 초대되었기에 자칫 뚜렷이 드러나고자 하는 충동은 더 클 수밖에 없다. 마치 중세의 카테드랄이 하얀 색이고 싶었던 것처럼, 신사대로의 랜드마크이고 싶은 백색의 유혹은 떨치기 힘들었을 것이다. 그럼에도 불구하고 강력하게 클로즈업 된 외피를 따라 이 타워를 올려다보며 갖는 큰 기대감이, 편한한 현관홀을 지나고 납작하게 반복되는 실내 공간으로 들어가면서 급감하게 된다. 이것은 아직 건축물이 미완성이어서 인가, 아니면 최대의 임대면적을 획득해야만 하는 상업 오피스 프로그램의 태생적 한계 탓인가. 이상봉타워는 강남의 대로변에서 각종 패션 컬렉션을 주도하고 아트 갤러리를 비롯한 복합 문화공간으로 현란한 프로그램을 구성하려한 건축물이다. 하지만 결과물로 볼 때는 중소 상업건물들 사이에서 몽유도원도의 피상적 해석에 숨어서 가로미관에 낯설음만 추구하는 건축적 태도를 보이고 있는 듯하다. 단면상으로 살펴 볼 때에도 안팎 볼륨 간의 교류가 기대에 미치지 못하고 심히 미약하다. 보란 듯이 차별을 시도하는 정면 파사드에 비해 측면에서 창문을 규정하는 방식 즉 창의 모서리만 둥굴리는 것도 내부의 기능과는 무관한 해결책으로 보인다. 이는 도시 컨텍스트에 걸맞게 입면의 대비를 의도한 것일 수도 있겠으나 그리 조화롭지 않다. 이는 마치 성형을 하고 화장을 뚜렷하게 시도했으나 오히려 근육이 굳어버려 생기를 잃고 생경스러운 표정으로 고착돼 가는 현상과 같다. 이미 현대 도시가 더 이상 저항할 수 없는 인스탄트 시티라며 표피건축의 순간성에 순순히 편승해야 하는지 의문이다. 역동적이고 복합적인 프로그램과 풍요로운 공간의 깊이를 여전히 필요로 하는 도시 서울에서, 대지와 자연이 건축물과 긴밀하게 연계되기를 기대하고 부단히 요청되던 대지건축 개념의 본질에 비해, 이상봉타워는 낯설은 표피 드러내기 작업에만 열중한 나머지 행여 때늦은 바로크적 경향성에 갇혀 버리는 것은 아닌가.

    언제부터인가 낯설게 던져지는 표피적 유희가 운생동의 건축작업을 통해서 빈번히 나타나고 있다. 그들은 누구보다도 실험적 건축개념과 색다른 디자인을 시도하려는 예술집단이고 싶어 한다. 이는 새로운 것을 향해 도전한다는 의미에서 분명히 긍정성을 띠고 있다. 아직 실패의 우려도 가지고 있지만, 그러한 불이익을 감내하면서까지 시대를 앞질러 개척해 보려는 아방가르드적 태도를 견지해 왔다고도 볼 수 있다. 하지만 건축가가 아직 젊든, 미래를 향해 도전적이든, 새로움의 충격을 지향하든 간에 건축 작업에서의 영속적 가치와 미학적 고상함, 그리고 시대를 막론하고 요청되는 건축 생산기술의 진보와 감동적 공간을 창조하기 위해서 다시금 자세를 고쳐 잡을 필요가 있다. 운생동이여, 이제는 랜드 스케이프에서 비롯되는 깊숙한 스킨 스케이프를 향하여 달려가라.

Gangnam after ten-something years of having DNA very similar to Gallery Yeh, is also trying hard to break the existing urban perspective and reach a new architectural and urban value. However, having been invited to a much bigger road than the time of Gallery Yeh, which had to lay in a narrow alley, it is inevitable that the urge to be clearly revealed is bigger. Just as medieval cathedrals wanted to be white, the temptation of becoming a white landmark of Sinsa-daero would have been hard to resist. Nevertheless, the great expectation that builds up while looking up at this tower along a strong close-up of the skin, drops sharply as you pass through a flat entrance hall to a flat, repetitive interior space. Is this because the architecture is incomplete or is it due to the inherent limitations of the commercial office program that must acquire the largest lease area? Lee Sang-bong Tower is a building that leads various fashion collections in the main street of Gangnam and that tried to construct a brilliant program as a multicultural space including an art gallery. However, based on the output, it seems to be taking on an architectural attitude of pursuing the unfamiliarity of the landscape, hidden by the superficial interpretation of the Mong-yoo-do-won-do between small/medium commercial buildings. Even when we look at the cross section, the interchange between the inside and outside volume is not as expected and is very weak. Compared to the frontal facade, which attempts to proudly distinguish itself, the way of defining the window on the side, i.e., rounding the corner of windows, seems to be a solution that is not related to the internal function. This may be intended as a contrast to the urban context, but it is not so harmonious. It is like a phenomenon where one underwent plastic surgery and put on bold makeup, but rather, ended up with hardened muscles, causing one to lose vitality and become someone with a stiff expression. I question whether we should passively get in the bandwagon of the momentary nature of skin architecture, saying that today's city is already an instant city that can no longer be resisted. In the city of Seoul, which still needs dynamic and complex programs and a rich depth of space, in contrast to the essence of the land architecture concept, which was anticipated and constantly requested of to closely link earth, nature and architecture, I wonder if Lee Sang-bong Tower is so focused on revealing its unfamiliar skin, trapping it in a belated Baroque inclination.

    At some point, play with unfamiliar skin is often seen in the architectural work of UnSangDong. More than anyone else, they strive to be more of an art group experimenting with experimental architectural concepts and unusual designs. Without a doubt, this is positive in the sense of challenging themselves towards something new. Although there is yet the possibility of failure, one could say that they have adhered to their avant-garde attitude to pioneer ahead of the era, despite having to endure such disadvantage. But whether an architect is still young, trying for the future, or aiming for the impact of something new, we need to fix our attitudes in order to create lasting value and aesthetic elegance in architectural work, to create a touching space, and for the advancement of architectural production technology demanded no matter what era we are in. UnSangDong, now run towards the innermost skinscape that begins from landscape.

Seoul Louis Vuitton Maison

서울 루이비통 메종

Location: Cheongdam, Seoul, Republic of Korea **Use:** Commercial **Site Area:** 938 m² **Building Area:** 599 m² **Building Coverage ratio:** 63.85% **Gross Area:** 3,841 m² **Structure:** Steel **Height:** 24m **Photographer:** Jaekyeong Kim(Model)

떠있는 커튼(floating curtain)은 새로운 인터페이스(interface)로 루이뷔통의 새로운 세계로 들어가는 길이다. 커튼을 올리고 새로운 항해를 꿈꾼다.

    루이비통 세계로 들어가는 새로운 인터페이스의 건축물을 제안한다. 고객과 루이뷔통 간의 의사소통 관계를 형성하는 플래그쉽(flagship) 센터로서의 건축물을 제안한다. 이 새로운 인터페이스는 루이뷔통의 절대적 욕망과 꿈, 역사, 장인 정신의 경계를 따라 이동한다.

    이 새로운 인터페이스는 마스터 장인과 고객의 관계, 기술과 예술의 경계, 상업과 예술의 경계를 실현하는 건축물이다. 쌍방향 브랜드 경험을 창조함으로써 고객들은 루이뷔통을 새기고, 루이뷔통은 고객의 마음을 이끄는 인터페이스를 형성하게 된다.

The floating curtain is new interface and entering into the new world of Louis Vuitton. Raise the curtain and dream a new voyage.

    An architecture as a new interface entering the world of Louis Vuitton is proposed. An architecture as flagship center which creates the relationship of communication between customers and Louis Vuitton is proposed. The new interface travels through the boundary of searching and entering the absolute desire and dreams, history and craftsmanship of Louis Vuitton.

    This new interface is an architecture realizing relationship of master craftsman and clients, boundary of technique and art, and the boundary of commerce and art. Creating bilateral Brand experience leads clients to engrave Louis Vuitton and it formulates interface attracting clients' heart.

Seoul Architecture Biennale Pavilion
프레서울건축비엔날레 파빌리온

Photographer: Junhwan Yoon, Unsangdong

세종대로는 조선, 대한제국, 현재의 대한민국에 이르기까지 역사가 살아 있는 심장부이다. 조선의 서학당길과 대한제국의 신작로가 만나는 (구)국세청별관을 허물어낸 장소에서 프레서울건축비엔날레를 개최하는 것은 한국 현대건축의 또다른 전기를 마련하는 계기가 된다. 시청, 덕수궁, 성공회 사이에 놓여진 대지는 추후에 시민에게 열려진 공원으로 활용될 계획이다. 잠시 임시적으로 사용될 프레서울건축비엔날레 파빌리온은 짧은 시간에 전시공간을 구축해야하는 어려움과 한계를 요구받았다. 기존의 대지를 그대로 활용한 가설건축물 형식의 공간을 제안한다. 일정한 간격의 철골기둥과 지형적 형태의 지붕을 결합하여, 전체가 하나의 공간으로 열려진 건축전시공간을 구성한다. 기둥의 숲으로 이루어진 유니버설의 전시공간은 미디어 모니터, 가벽, 전시박스에 의해서 채워진다. 다양한 전시의 가변성, 하나의 공간으로 연속된 통일성을 만들어 내는 개념을 담는다. 전체가 하나의 공간이지만 3개의 레이어가 분리된 경사형태의 지붕을 통해서, 틈사이로 떨어지는 자연광, 서로 다른 높이의 천정으로 구성된 입체적인 전시공간을 구성한다. 도시에 임시적으로 만들어지는 산수형태의 도시지붕을 제안하는 것이다.

Sejong Avenue is the center of history from Chosun, Korean Empire to the present Korea. Holding the Seoul Architecture Pre-Biennale at the place where the old annex to the National Tax Service building existed, where the Seohakdang-gil from the Chosun period and a new road from the Korean Empire meet is another opportunity for Korea's modern architecture. The target site, located among the City Hall, Deoksugung Palace, and the Anglican Church of Korea, will be used as a park open to citizens in the future. The pavilion at the Seoul Architecture Pre-Biennale, which will be in temporary use, has the difficulties and limitations of establishing the exhibition space in a short time.
We propose that the space for the temporary building utilize the existing land as it is. By combining steel columns with regularly spaced roofs sections and a topographical roof form, the entire building becomes an open exhibition space. The universal exhibition space, made up of wooden pillars, is filled with media monitors, temporary walls, and display boxes. It contains a variety of numerous exhibitions and the concept of creating consecutive unity in one space. It is a single space, but it forms a three-dimensional exhibition area consisting of natural light falling through gaps and ceilings of different heights which are composed of a sloped roof with three separated layers. This urban roof suggests the shape of a mountain that is temporarily relocated in the city.

For the Universalization of Gi-un-saeng-dong
기운생동 (氣韻生動) 감각의 세계적 보편화를 위하여

Hwang Yi (Department of Architecture, Ajou University)
이 황 (아주대학교 건축학과)

운생동 건축사사무소(이하 운생동) 은 매우 특별한 건축사무소이다. 장윤규, 신창훈 소장(이하 장윤규, 신창훈)과 개인적인 인연은 없으나, 대학원 시절 그들이 설계한 장소 (서울대학교 공과대학 39동) 에 첫 입주하여 학업을 이어갈 수 있었던 행운아 중 한 명이었다. 성숙하지 못한 한국 사회의 혼란스런 현실에도 불구하고, 여러 해의 노력이 뜻깊은 결실을 맺는 이번 전시를 통하여 소개될 많은 작품들을 보며, 새로운 건축과 공간구축을 향한 끊임없는 비전과 그리고 그것을 실현하기 위해 노력한 스탭들의 지단한 분투과 성실한 도전에 아낌없는 찬사를 보내는 바이다.

장윤규는 일찍이 "복합체 201 (장윤규, 2005)" 이라는 책을 통해 자신만의 작업 방향을 정리하여 소개 한 바 있으며, 초기 시절 스스로 밝혔듯이 (이은경, 김혁준, 2004), 들뢰즈의 "-되기"/"-하기" 등을 비롯한 다소 추상적 혹은 개념적인 텍스트로부터 선택적 영감을 얻고, 그것으로부터 공간적인 상상력을 구체화시키는 방향으로 일관된 작업을 진행하고 있다. 근작인 미동 전자 사옥 (2016), 이상봉 타워 (2018)에 각각 "퓨처리즘 그리드 (Futurism Grid)", "몽유도원도" 라는 별칭을 붙였듯이 (장윤규, 2018), 언어적 유추를 통한 물리적 형식의

There is something special about the architectural office UnSangDong Architects Cooperation (USDSpace). Though I do not personally know principals Jang Yoon-Gyoo (Jang) and Shin Chang-Hoon (Shin), I was lucky enough to be one of the first occupants of Seoul National University College of Engineering Building 39, a place that they designed, when I was studying for my Master's. This exhibition presents many works, which are the worthwhile fruits of many years of hard work, all despite the unstable reality of Korea's still maturing society. I extend my heartfelt compliments to the unwavering vision towards constructing new architecture and space, and to the strenuous and sincere efforts of the staff who worked hard to realize it.

Jang Yoon-Gyoo has introduced the direction of his work early on through a book titled Compound Body (Jang Yoon-Gyoo, 2005). As he said for himself at the beginning of his career (Lee Eun-kyung, Kim Hyeok-Jun, 2004), he gets selective inspiration from somewhat abstract or conceptual texts, such as Deleuze's "becoming" or "doing", and is working in a consistent direction to embody spatial imagination from it. Jang has nicknamed his latest works, Midong Electronics & Telecommunication Headquarters (2016) and Lee Sang-bong Tower (2018) "Futurism Grid" and "Mong-yoo-do-won-do" (Jang Yoon-Gyoo, 2018). In line with the theme "imagination object (Lee Young-beom)", which is the

변형에 대한 지속적인 관심은 그가 추구하는 새로운 형태 생성의 결과이자 목적인 "상상체 (이영범, 2013)"라는 주제와 맞물려, 고정된 형태를 넘어 (지어진 이후에도) 새로운 건축으로 재생산/해석되길 바라는 노력으로 비춰진다.

천의영 교수 (2014)는 운생동의 작업이 한국 건축의 "특개성 (singularity)"을 잘 드러내는 것이라 평한 적이 있으며, 송하엽 교수 (2018) 또한 이에 동의하고 있다. 건축가들에게 결코 호의적이지 않은 어려운 한국 환경에도 불구하고 그들만이 보유한 감수성과 감각이 갖는 건축적 혁신성을 높게 평가한 것이다. 운생동의 존재감은 최근 이탈리아의 「The Plan Awards 2018」을 비롯한 다수의 해외 수상, 그리고 국제 작품집 출간(Unsangdong Architects, 2017) 등을 통해 이미 잘 드러난 바 있으며, 다수의 협업을 통한 탄탄한 설계 조직 또한 세계적 건축 사무소로 거듭날 수 있음을 증명하고 있다. 그럼에도 불구하고, 혹은 그러하기 때문에 더욱, 이번 전시와 앞으로의 작업을 통해 그들의 독특함이 이루어낼 성취에 대해 더 많은 기대를 하는 것이 무리한 개인적 견해만은 아니라고 생각한다. 장윤규, 신창훈이 갖고 있는 "기운생동 (氣韻生動)"의 감각이 빠르게 변화하는 세계 현대 건축의 흐름과 사회, 문화적 컨텍스트 속에서 보편성을 갖고 이론화, 확장될 때, 그들의 작업이 비로소 우리가 기대하는 국제적인 건축가의 반열에 오르게되리라 믿기 때문이다.

다른 전문직들과는 다르게 건축가들은 자신만의 원칙과 그 원칙에 의한 결과로 다시 원칙의 타당성을 평가 받아야하는 모순적 상황에 늘 직면하게 된다. 제레미 틸 교수 (2012)가 지적하듯, 적지 않은 현대 건축가들이 이런 위험을 피하기 위해 근대 건축의 그림자가 주는 손쉬운 구원의 길에 빠지거나, 혹은 포스트 모더니즘이란 이름의 일탈로 후퇴해 버렸다. 이번 전시를 맞아 운생동의 노력이 보편적 재생산과정을 거치기 위해 장윤규가 그의 건축을 통해 지향하는 관점과 질문의 예리함과 일관성, 그리고 그 대답을 실천하기 위해 사용하는 형태, 기술적 도구들의 적절성에 대한 면밀한 고찰과 건설적 비판이 선행되어야 하는 이유이다.

한국 현대 도시, 건축의 혼잡함이 근대적 도시성장의 파편 혹은 변종이라고 본다면 (임석재, 2013), "복합체, 다양체 (장윤규, 2014)"라는 키워드를 통해 건축가로서 그가 던지는(상상력을 투영하는) 질문의 대상은 현실의 도시와 우리의 거주 방식이라는 보편적 대상이다. "리좀 (rhizome)", "코라(chora)" 등을 포함한 포스트 모더니즘의 감각적 어휘들의 차용 (장윤규, 2014; 장윤규, 2018) 을 비롯해, 미동전자 사옥 (2016)을 설명하면서 사용한 "데카르트적 사고체계에 대한 공격 (장윤규, 2018)" 이란 표현을 보면, 장윤규는 우리 사회가 근대의 산물이며 현대의 복잡성도 그 연장선에 있다고 하는 건축가들의 기본적인 인식 (Nesbitt, 1996)을 공유하는 듯 보인다.

result and purpose of creating a new form that he pursues, the continued interest he shows in the transformation of physical forms through linguistic analogy can be reflected as an effort for his work to go beyond a fixed shape and be reproduced/understood as a new building (even after construction is completed).

Professor Chun Eui-Young (2014) has commented of USDSpace's work as revealing the singularity of Korean architecture, a statement supported by Professor Song Hayub (2018). This is to say that they highly appreciate the architectural innovativeness of their sensibility and sense despite the difficult Korean environment, which is definitely not favorable to architects. USDSpace's presence has increased over recent years through numerous international awards including the "The Plan Awards 2018" in Italy and the international publication of a collection of works (UnSangDong Architects, 2017). A sound design organization created through multiple collaborations is also proof of the possibility of being reborn as a global architectural office. Nonetheless, or all the more because of this, I think it is not an unreasonable, personal opinion to say that I have high expectations of their future accomplishments, which will be achieved by their uniqueness through this exhibition and future work. This is because I believe that when Jang and Shin's sense of "gi-un-saeng-dong (qìyùn shēngdòng; vivid capture of rhythm or spirit)" universally theorizes and expands within the rapidly flowing global modern architecture and the social and cultural context, this is when their work will join the ranks of international architect.

Unlike other professions, architects are always confronted with contradictory situations in which the validity of their principles must be evaluated again by their own principles and the derived results. As Professor Jeremy Till (2012) points out, many modern architects have fallen into the path of easy salvation provided by the shadows of modern architecture to escape such dangers, or retreated into the deviance in the name of postmodernism. In order to undergo the process of universal reproduction in this exhibition, this is the reason why meticulous contemplation and constructive criticism on the sharpness and consistency of the perspective and questions that Jang aims for in his architecture, and appropriateness of the forms and technical tools used to put the answers into practice should be preceded. If the confusion of modern cities and architecture in Korea is regarded as a fragment or variant of modern urban growth (Lim Seok-jae, 2013), the object of the question he throws (imagination he projects) as an architect through the keywords "compound body, manifold body" (Jang Yoon-gyoo, 2014) is a ubiquitous object of the city in reality and the way we live. From the use of sensuous postmodern vocabularies such as "rhizome" and "chora" (Jang Yoon-Gyoo, 2014; Jang Yoon-Gyoo, 2018) to the use of the expression "attack on the Cartesian thinking system (Jang Yoon-kyu, 2018)" in describing Midong Electronics Headquarters (2016), Jang seems to share the basic perception of architects that our society is the product of modern times and that the complexity of contemporary times is an

이러한 보편적 메타 인식을 통한 각론적 해결들의 타당성을 검토하기 위해 우리는 건축가 함인선 교수 (2014)가 요약하는 근대성 (modernity)을 결정짓는 세가지 특징- 공공성, 기술, 자본-에 대해 주목할 필요가 있다. 제레미 틸(2012)이 새로운 건축 형태의 등장을 순환식 엘리베이터인 파터노스터 (paternoster) 에 빗대어 비판하듯, 세 가지 모두 변화하지 않으면 완전히 새로운 건축은 결코 실현되거나 주장될 수 없기 때문이다. 운생동의 "스킨 스케이프 (Skin scape)", "인간이 동물 되기 (Being animal)", "클립 시티 (Clip city)" 라는 차별화된 주제어들은 (장윤규, 2005; Unsangdong Architects, 2017), 근대성에 대한 작가의 비판을 바탕으로 하기에 역설적으로 매우 보편적이다. 건축 생산에 직접적으로 작용하는 실행어들이 만들어낸 여러 작품들- KT&G 수원 복합 문화센터 당선작 (flying city), Kring 복합 문화공간 (2008), 성동구 문화복지센터 (2012), 국립 무형 유산원 (2013) 등-은 큰 스케일의 공간과 복잡한 구조적 제스처가 만나 건축기능-프로그램의 재조직과 복합을 주장함으로서 다른 건물이 보여주지 못한 건축의 "도시적 공공성"-근대성의 첫번째 특징-에 대한 새로운 대답을 성공적으로 추구하고 있다.

여러 건축인들이 공감하듯이 (천의영, 2014; 송하엽, 2018) 운생동의 건축적 해답에서 공통적으로 발견되는 하나의 큰 특징은 과감한 형태적 복잡성이다. 그 복잡성은 크게 보아, 1) 건축과 물질의 대지와 중력의 종속에 대한 근본적 의문 (광주 국립아시아문화의전당 계획안 (2006), 2012 여수 엑스포 파빌리온 계획안 (2010)), 2) 건물의 구조와 외피의 이분성에 대한 질문 (예화랑 (2005), 오션어스그룹 해운대 사옥 (2014))에서 직접적으로 비롯된 것으로 볼 수 있다. 균질함을 담보로한 확장성, 측정되고 나누어져야하는 근대적 공간을 타겟으로 이어져온 담론을 건축 구조와 물질의 변형에 대한 관심으로 전환 시킨 것은 운생동을 이전의 한국 건축가들과 차별화 시킨 큰 원동력임에 틀림 없다. 그러나, 운생동이 추구하는 독창적 건축 형태의 보편 타당성은 그것이 새로운 공간적 체험을 제시하는가에 대한 검증 뿐만 아니라, 근대성을 벗어나기 위한 두번째 핵심 과제인 "기술적 혁신"-재료, 디자인 프로세스, 생산 방식 등의 변화-을 담보로 해야만 얻어질 수 있음을 명확히 해야 한다. 로버트 벤추리 (Venturi, 1996) 가 지적하듯이 형태적 복잡성은, 장윤규의 감각적 붓 스케치에서 염려되는 그러나 그의 의도와는 반드시 다를 것인, 자칫 회화적 형태에 대한 의지의 표현 (self-willed expression of picturesqueness)으로 잘못 이해되버릴 수 있기 때문이다.

우리는 건축가들이 앞서 말한 근대적 특징의 극복을 바탕으로한 보편성 획득에 노력하고 있고 그것을 계기로 세계적 명성을 쌓아 왔음을 알고 있다. Herzog & De Meuron은 Prada Aoyama Epicenter (ELcroquis editorial, 2006) 에서 구조와 외피가 완벽히 통합된 공간을 제시한 바 있고, OMA 는 서울의

extension of this (Nesbitt, 1996).

In order to examine the feasibility of these theoretical solutions through this universal meta-awareness, we need to pay attention to three characteristics that determine modernity - publicness, technology, and capital - summarized by architect and professor Ham In-sun (2014). As Jeremy Till criticizes the emergence of a new architectural form by comparing it to a paternoster (2012), completely new architecture can never be realized or claimed unless all three are changed. USDSpace's differentiated topic words, "Skin Scape", "Humans Becoming Animals", and "Clip City" (Jang Yoon-Gyoo, 2005; UnsangDong Architects) are, paradoxically, too universal to be based on the artist's criticism of modernity. Various works created by practical words that directly affect architectural production, such as the winning entry of KT&G Suwon Culture Complex (Flying City), Kring Kumho Culture Complex (2008), Seongdong Cultural Welfare Center (2012), National Intangible Heritage Center (2013) and other works successfully pursue a new solution to the "urban publicness" of architecture--the first characteristic of modernity--that other buildings have not been able to show. This is acheived by asserting the reorganization and complexity of architectural functions and programs through the contact of large scale spaces and complex structural gestures.

As many architects agree (Chun Eui-Young, 2014; Song Hayub, 2018), one major feature commonly found in USDSpace's architectural solutions is bold, morphological complexity. Largely, this complexity can be seen as being directly derived from the following: 1) The fundamental question about the subordination of architecture and materials to earth and gravity (Proposal for Asia Culture Center in Gwangju (2006), Proposal for 2012 Yeosu Expo Pavilion Project (2010)), and 2) the question about separation of a building's structure and skin (Gallery Yeh (2005); Ocean Us Haeundae Office (2014)). Expanding while preserving homogeneity, and converting discourse, which until now has been aimed at modern space that has to be measured and divided, into an interest in architectural structure and material transformation is no doubt a big driving force that distinguishes USDSpace from previous Korean architects. However, it should be made clear that the universal feasibility of the original architectural style pursued by USDSpace is not only a verification of whether it presents a new spatial experience, but that it can only be obtained by collateralization of the second key task to escape modernity: "technological innovation" i.e., change in materials, design process, production method. As Robert Venturi (Venturi, 1996) points out, morphological complexity can be misunderstood as a self-willed expression of picturesqueness, which is to be wary of, but definitely not intended in Jang's sensuous brush sketches.

We are aware that architects have been trying to acquire universality based on the overcoming of modern features mentioned above and have built a global reputation as a result. Herzog & De Meuron presented a space in which the structure and skin are fully integrated in

프라다 파빌리온 계획 (Prada Transformer, 2009) 을 통해 중력에 저항하는 공간적 장치에 시간적 변형을 통해 건축 프로그램을 복합적으로 이식할 수 있음을 증명하였다. 운생동의 매력적인 형태 감각에도 불구하고, 해운대 사옥, 미동 전자 사옥, Kring 복합 문화 공간등에서 아쉽게 드러나는 완전히 봉합되지 못한 구조-스킨-공간의 통합, 그리고 Rooftecture (E+ Green Home) 에서 발견되는 기술적 사실과 형태적 통일의 충돌에 대한 몇 가지 모순은 그들이 앞으로 기술 혁신을 바탕으로한 보편성의 획득이라는 두 번째 과제에 절실하게 도전해야하는 이유를 드러낸다.

    운생동만의 창의적 상상력을 뒷받침할 완성도에 대한 아쉬움은, "기운생동의 세계화"를 가로막는 이유에 대해 두 가지 층위의 또 다른 논의를 요구 한다. 좁게는 신승수 소장 (신승수, 2006) 이 지적하듯이 뛰어난 디자인을 완벽히 실현할 기술력과 컨설팅의 부재에 기인한 척박한 우리 건축 현실에 대한 자기 비판이며, 보다 넓게는 건축에 대해 우리 건축인과 사회가 갖는 지각은 여전히 근대에 머물러 있지 않나하는 반성에 관한 문제이다. 디자인과 엔지니어링의 간극, 실무와 이론의 간극은 종종 건축에 대한 이분적인 평가 오류-완전히 기술적인 측면에서 평가하거나 혹은 완전히 예술적인 차원에서만 인식하는 것-로 나타나 건축가에게 건전한 피드백을 제공하지 못하는 주요한 원인이 되고 있다. 또한, 건축가들 조차 건축의 3요소-건축, 건축가, 건물-가 근원적으로 서로 다른 체계에 놓일 수 밖에 없다는 상황을 간과하고, 디자인의 의지와 보편적 개념의 획득을 건축적 자율성의 문제로만 파악하는 우를 범하기도 한다.

    건물이 완성되는 순간 사회에서 소비되는 생산물임을 인정하지 않고, 건축적 사고의 발현으로서만 존재하기를 바라는 한, 존 듀이 (John Dewey)와 쿠마 켄고 (Kengo Kuma)가 말하듯 현대 건축의 담론은 빠르게 변화하는 현대 사회에서 쉽게 보편화되기 힘든 예술적, 철학적 텍스트들로 어지럽혀 지거나 미디어와 자본에 종속될 수 밖에 없다 (제레미 틸, 2012; 쿠마 켄고, 2009). 한편, 전문직으로서의 건축가라는 독립적 영역자체가 근대의 산물이듯 그 안에서 얻어진 해답조차도 모더니즘의 한 부분이기에 파편적이고 제한적일 수 밖에 없다는 한계도 우리는 인식해야만 한다. 그러나 여러 어려운 상황에도 불구하고, 장윤규, 신창훈 소장이 왕성하게 생산해온 작품들과 글과 이미지들은 후학과 동료 건축가들에게 지대한 영향을 주었고 그 영향력은 지금까지도 지속, 증대 되고 있다. 그들의 대담한 제스처와 열정, 창의적인 감각의 결과는 비록 아직은 미완의 해답일 지언정, 그것을 통해 우리가 얻어야할 교훈까지도 제시한다고 해석하고 싶다. 운생동이 "-되기"와 "복합" 이라는 키워드를 통해서 치열하게 탐구하고 실천해온, 새롭게 건축을 생각하는 방법, 새롭게 구축하는 방법의 문제를 건물이 생산되는

the Prada Aoyama Epicenter (EL Croquis Editorial, 2006), and OMA demonstrated through the plan for Prada Transformer in Seoul (2009) that it is possible to complexly transplant an architectural program through temporal transformation in a spatial device that resists gravity. Despite USDSpace's attractive sense of form, the not-completely-seamless integration of structure, skin and space found in buildings such as the Haeundae Office Building, Midong Electronics Headquarters and Kring Kumho Culture Complex, and some contradictions of technical facts and the conflict of morphological unification found in Rooftecture (E + Green Home) reveal why they are hereafter pressed to challenge themselves against the second task of acquiring universality based on technological innovation.

    The regrettable lack of completeness to support USDSpace's creative imagination requires another discussion of two different levels as to what is preventing "the globalization of gi-un-saeng-dong". Narrowly, as Shin Seung-soo (Shin Seung-soo, 2006) pointed out, it is self-criticism of our architectural reality, which is barren due to lack of technical expertise and consulting required for complete realization of outstanding design. More broadly, it is an introspective issue of whether the perception on architecture that we architects and society holds is still stuck in the modern age. The gap between design and engineering, and the gap between practice and theory, is often a major cause of failure to provide sound feedback to architects. It can be in the form of a binary evaluation error of architecture--either entirely technical or entirely artistic. In addition, even architects overlook the situation that the three elements of architecture--architecture, architect and building--can not but be on fundamentally different systems, and make the mistake of recognizing the will of design and the acquisition of universal concepts solely as problems of architectural autonomy.

    As long as we do not acknowledge that a building is a product consumed in society at the moment of completion, and as long as we wish for it to exist only as a manifestation of architectural thinking, as John Dewey and Kengo Kuma say, the discourse of contemporary architecture is bound to be confused with the artistic and philosophical texts that are difficult to be universalized in a rapidly changing modern society or to be subordinate to media and capital (Jeremy Till, 2012; Kengo Kuma, 2009). Meanwhile, as the independent domain of an architect as a profession is a product of the modern age, we must recognize that even the answer derived from it is a part of modernism, and thus is fragmented and limited. Despite the difficult situation, however, the works, texts and images that Jang and Shin have produced vigorously have had a profound impact on their juniors and fellow architects, which continues and is steadily increasing even today. Though the results of their bold gestures, enthusiasm and creative sensations are yet an incomplete solution, I would like to interpret this as the provision of lessons we should learn. Through the keywords "becoming" and "composite", USDSpace has thoroughly explored and practiced new ways of thinking about and

물질적, 기술적 과정과 연구에 더 세밀하게 초점을 두어 나아간다면, 그들이 마땅히 지켜가야할 한국 건축에서의 선지적 존재감을 세계를 선도하는 무대로 한 단계 더 끌어올리는 계기가 되지 않을까 생각한다.

constructing architecture, and the material and technical process and research on which buildings are produced. If they continue forward with a greater focus on this process and research, I think that it would be an opportunity to raise its leading presence in Korean architecture, which they deserve to keep, up another level to a stage one step ahead of the world.

참고 문헌 (References)
1 장윤규, 복합체 201 COMPOUND BODY, 서울: 간향 미디어랩, 2005.
2 이은경, 김혁준 (편집), 2004. 전인호와 장윤규가 말하는 건축 그리고 그 무엇. SPACE 438, 90-95. (Lee, E. K. and Kim, H. J., 2004. The architecture and some more spoken by jun, in ho and jang, yoon gyoo, SPACE 438, 90-95.)
3 장윤규, 2018. 논리적 오류의 풍경. SPACE 609, 38-41. (Jang, Y. G., 2018. The Landscape of Petitio Principii. SPACE 609, 38-41.)
4 이영범, 2013. 복합체에서 사회적 상상체로의 전이. C3 KOREA 346, 144-149. (Reigh, Y. B., Transformation of Compound Body to Social Imaginative Body. C3 KOREA 346, 144-149.)
5 천의영, 2014. 한국건축의 특개성과 운생동의 근작. SPACE 560, 52-59. (Chun, E. Y., 2014. The singularity of Korean architecture and Unsangdong architects' latest work. SPACE 560, 52-59.)
6 송하엽, 2018. 도시추상화: 편린이 반복되는 추상화된 표면. SPACE 609, 56-57. (Song, H., 2018. Urban Abstract Canvas: The Abstract Surface with Repetitive Partitions. SPACE 609, 56-57.)
7 Unsangdong Architects, 2017. Compound body. l'ARCA International, S.A.M.M.D.O., MC, Monaco.
8 제레미 틸, 불완전한 건축. 이황 (옮김), 서울: SPACETIME, 2012. (Till, J. Architecture Depends. Trans. Yi, H., Seoul: SPACETIME, 2012.)
9 임석재, 한국 현대건축의 지평. 서울: 인물과 사상사, 2013.
10 장윤규, 2014. 다양체의 지도. SPACE 560, 36-51. (Jang, Y. G., 2014. The map of manifold. SPACE 560, 36-51.)
11 Nesbitt, K., eds., Theorizing A New Agenda for Architecture: An anthology of architectural theory 1965-1995. New York: Princeton Architectural Press, 1996.
12 함인선, 정의와 비용 그리고 도시와 건축: 근대 건축으로 한국사회를 읽다. 서울: 마티, 2014.
13 Venturi, R., Complexity and contradiction in architecture: selections from a forthcoming book, In Nesbitt, K., eds., Theorizing A New Agenda for Architecture: An anthology of architectural theory 1965-1995. New York: Princeton Architectural Press, 1996.
14 ELcroquis editorial, Herzog & De Meuron 2002/2006, Madrid: ELcroquis publisher, 2006.
15 신승수, 2006. 상상력의 계획. C3 KOREA 0601, 158-159. (Shin, S. S., 2006. Planning Imagination. C3 KOREA 257, 158-159.)
16 Prada Transformer, 2009. Web, Accessed, Aug. 23. 2018. Retrieved from http://oma.eu/projects/prada-transformer.
17 쿠마켄코, 약한 건축. 임태희 (옮김), 서울: 디자인 하우스, 2009.

Jeonju Intangible Cultural Heritage Hall
전주 무형문화유산의 전당

Location: 896-1 Dongsuhhak-dong Wansan-gu, Jeonju, Republic of Korea **Use:** cultural facility **Site area:** 63,020.00m² **Building area:** 13,727m² **Gross floor area:** 33,174.64m² **Building coverage:** 21.78% **Floor space index:** 37.51% **Building scope:** B1 ~ 4F **Structure:** Steel + Reinforced Concrete **Photographer:** Junhwan Yoon, 토문엔지니어링 제공

미래적이고 모던한 건축을 통하여 전통의 공간을 재구성해내야 하는 '무형문화유산전당'은 운생동건축에게는 새로운 도전의 분야라는 생각을 가지게 되었다. 하나의 장소 속에는 과거, 현재, 미래가 응집된다. 마치 고고학의 편린처럼 존재하는 장소, 혹은 공간은 우리들 서로에게 공존의 의미를 주며, 서로 공명하며 수많은 언어를 주고 받기를 원한다. 무형문화유산전당 프로젝트을 통해서 시간의 개념을 넘나드는 새로운 고고학적인 틀을 설정하기를 원했다. 기존의 사이트의 보존과 개발의 부분을 넘나드는 하나의 경계로 '무형의 틀 Intangible Frame'를 설정한다. 숲이 지녔던 땅의 기억과 틀 안에서 만들어질 내일의 기억을 함께 보존하는 틀이다. 한국 전통건축의 회랑의 공간구조를 현대적으로 변형하여 '무형의 틀'을 구성한다. 전통건축에서의 회랑은 영역을 구분하면서도 건물과 건물을 서로 이어주는 경계의 교묘한 틀의 역할을 수행하였다. 이러한 개념을 '네트워크적 회랑Network Corridor'의 개념으로 더욱 강화된 시스템으로 변환시킨다. 무형문화유산전당의 프로그램인 9개 센터와 외부 이벤트의 마당은 회랑 복도와 데크를 통해서 하나의 통합체가 된다. 필요한 기능의 박스는 회랑으로 구성된 틀속에서 재배열된다.

The aim of Intangible cultural heritage hall is re-organizing traditional space to futuristic and modern architecture which is a new challenge to us. Past, present and future are condensed into one place. The space exists like a part of archaeology. Or maybe it gives a meaning of coexistence to us and wants to resonate giving and taking numerous languages. We want to set up new archaeological frame covering the concept of time. Intangible Frame, we propose, is in between development and preservation of existing cultural heritage. This is the frame preserving the memory of tomorrow which will be made from memories of the earth and the forest. Korean traditional corridors transform to contemporary intangible frame. In traditional architecture, the corridor divides territory and connects buildings with buildings. We emphasize this function as a network corridor. 9 centers and a outdoor event garden are compounded through the aisle of the corridor and the deck. Boxes of necessary functions are re-arranged through the frame consisted of the corridor.

SK Networks Gangnam Office
SK 네트웍스 강남 사옥

Location: Yeongdong-daero, Gangnam-gu, Seoul, Republic of Korea **Site area:** 8,267.10 m²
Gross floor area: 47,307.83 m² **Photographer:** Junhwan Yoon, Sergio Pirrone

본 프로젝트는 중구 명동에 흩어져 있는 SK네트웍스 사옥을 강남구 대치동에 통합사옥으로 신축하는 프로젝트이다. BOX의 공간과 형태 자체가 입면개념이 되며, 건축의 공간, 매스, 스킨의 일관된 디자인을 통합하였다. 디자인의 조율과 통합은 SK네트웍스의 일관된 기업철학을 추상화한다. 또한 정면의 비젼부는 SK네트웍스와 도시와 소통하는 상징적 아이콘으로 수많은 NETWORK들의 창조적 결합을 통해 "미래의 가능성"을 현실로 바꾸는 건축물. 이러한 비젼을 시스템적으로 구현하였다.

This project is to build a new headquarters for SK Networks in Daechi-dong, Gangnam-gu, because the offices are spread throughout in Myungdong, Jung-gu.
The space and form of the BOX itself becomes the concept of the facade and incorporates the consistent design of the space, mass, and skin of architecture. The coordination and integration of the design reflects the consistent corporate philosophy of SK Network's. In addition, the vision of the facade is a symbolic that SK Networks communicates with the city. Through the creative combination of numerous networks, we systematically implemented the architecture that transforms the "possibility of the future" into reality.

Shallow Sense, Deep Surface
얇은 감각, 깊은 표면

Youchang Jeon (Department of Architecture, Ajou University / Partner, aDlab+)
전유창 (아주대학교 건축학과, aDlab+ 공동대표)

시각을 넘어서

15세기 이후 서구의 시각 양식의 중심으로 유지되어온 원근법의 패러다임은 19세기 이후 시점의 다중성, 새로운 매체의 등장, 새로운 광학 기구들의 발명 그리고 사진, 영화와 같은 시각 테크널러지의 등장에 의해 변화하게 된다. 대도시의 발전과 소비공간의 출현은 인간의 시각적, 지각적 체험양상에 커다란 변화를 가져왔다. 기차나 자동차등 고속화된 이동수단은 우리의 속도감각을 변화시켰다. 짧은 시간 안에 빠른 속도로 이동하며 경험된 시각적 정보는 제한된 수용능력을 가진다. 변화된 속도감각은 동시에 공간감각의 변화를 초래했다. 이는 특히 아케이드 및 백화점과 같은 도시의 소비문화공간에서 보여지는 '유동화된 응시'를 통해 대표적으로 나타난다. 속도는 과도한 시각정보의 축적을 유발한다. 또한 움직임이란 조건에 기초하여 파노라마적 시각체험을 가능하게 한다. 우리의 일상을 둘러보면 시각적 자극을 주는 볼거리로 가득찬 공간으로 둘러싸여 있다 건물의 입면과 상점, 전시된 상품들, 거리의 광고물들, 상점의 간판들은 일시적이고, 우연적이며, 파편적인 요소에 의해 산만하고 분산적인 체험의 형식이 가능해진다. 이러한 시각 경험의 변화는 우리의 감각을 표면이 만드는 이미지에 집중하게 만든다. 유희적

Beyond Visuality

The paradigm of perspective, which has served as the center of the Western visual style since the 15th century, was changed by a multiplicity of viewpoints in the 19th century, the emergence of new media, the invention of new optical instruments, and appearance if visual technology such as photography and film. The development of metropolitan areas and the emergence of consumption spaces brought about great change in human visual and perceptual experiences. Speedy transportation, such as trains and cars has changed our sense of speed. Moving at a rapid rate in a short period of time illustrated that visual information has a limited capacity. The changed sense of speed also caused an alteration in the understanding of space. This is particularly evident in the "fluidized gaze" seen in urban consumption cultural spaces such as arcades and department stores. Speed causes an accumulation of excessive visual information. It also enables panoramic visual experiences based on movement. When we look around in our daily lives, we are surrounded by spaces full of visual stimuli. The facades of buildings and stores, displayed merchandise, street advertisements, and signboards form distracting and dispersed experiences through temporal, accidental, and fragmentary elements. This change in visual experience makes our senses more focused on the images the surface

표현에 의한 욕구는 건축 표면에 자극적인 이미지의 표출을 가속화한다. 표면이 만드는 건축의 외피는 도시공간의 경관을 구성하는 가장 기본적인 요소이다. 건축 외피에 표피적이고 피상적인 것의 중요성이 대두되면서 표면은 다양한 형식과 방법으로 만들어진다. 운생동 건물의 외피는 공간과 표면, 문화와 기술, 구조와 형식 등의 다양한 층위의 이슈들을 담아내는 점에서 구축의 서사를 만드는 중심에 있다. 최근작인 몽유도원도, 미동전자부터 그들의 대표작이라 불리우고 있는 크링까지 운생동에서 설계된 대부분의 작품은 현대 도시라는 배경에서 표면을 통해 도시와 상호 작용하는 이미지를 설계한다. 그것들은 현대의 이미지 수용방식과 표현방식의 변화를 대변한다. 표면이 만드는 이미지에 대한 관심은 외피의 포장술을 넘어 '깊이의 구축'이라는 텍토닉적 사고에 의해 확장된다. 깊이는 표면에 심층의 영역을 만든다. 그리고 표면은 그 시대에 필요한 새로운 감각적 체험과 일상의 관계를 재정의한다.

촉각, 깊이의 시작

운생동이 만들어 내는 건축은 시각적 감각 시스템을 최대한 활용하여 표면안에서 벌어지는 건축 체험을 끊임없이 조작하고 있다. 그들의 건축에서 2차원의 표면은 3차원 세계의 환영과 표현을 만드는 데 사용된다. 때로는 우리의 눈은 3차원의 이미지를 체험하기 위해 2차원의 표면을 탐색한다. 표면에서 발현되는 재료, 패턴, 스케일 그리고 깊이의 특질은 눈을 넘어 보다 촉각적인 감각에 의해 깊게 측정되고 체험된다.

　　　　독일의 저명한 미술사학자 리글(A. Riegl)은 이집트의 부조에서 나타나는 시선으로 더듬으며 가까이 보여지도록 만들어진 시각을 '광학적인것(l'optique)'과 구별하여 '눈으로도 만지는 것(l'haptique)'으로 표현하였다. 부조의 표면은 눈과 손을 가장 엄밀하게 결합한다. 또한 평평한 표면은 눈으로 하여금 촉각처럼 움직이도록 허용해 주고 더 나아가서 눈으로 만지는 기능을 부여한다고 하였다.[1] 촉각은 다감각적(multi-sensory) 경험으로 건축의 표피에 대한 체험과 감각의 변화를 부여한다. 촉각과 시각이 만나는 '만지는 눈'이라는 현대화된 표면의 촉각성은 운생동 건축의 일관된 특징인 감각의 자극에 반응하는 건축 표면의 생성과 표면처리 기법과 밀접히 연관되어 있다. 이상봉타워의 곡면 스트립(Curvd Strip)패턴 크리스탈 마운틴의 타공된 접힌 주름 (Perforated Fold), 패턴 스트럭츄럴 매쉬에서 보여지는 외곡된 구조(Distorted Structure) 패턴, 크링의 잘려진 원형(Cutted Cirle) 패턴 퓨처리즘 그리드의 사선(Diagonal grid) 패턴 등등은 재료, 색, 물성, 텍스쳐, 패턴 등의 요소를 촉각적인 방식으로 사용한다. 이러한 감각적 표면의 구현은 인간척도의 스케일에 의해 표면이 잘개 부서지는 형상으로 나타난다. 패턴의 반복과 변화는 그들의 건축이 추상적 감각의 구현 방식으로

creates. The desire for amusing expression accelerates the irritating images expressed on architectural surfaces. The outer skin of buildings that the surface creates is the most basic element, constituting the landscape of the urban space. As the importance of the superficial nature of architecture's outward look has been on the rise, the surface is being made in more various forms and methods. The outer skins of the buildings by UnSangDong are at the center of creating a narrative of construction, in terms of containing various issues on different levels such as space and surface, culture and technology, and structure and form. Most of the works designed by UnSangDong, from the recent "Mongyudowando," "Midong Electronics and Telecommunication Headquarters," and "Kring," the most representative work, are designed to interact with the modern city through the surface. These creations represent changes in modern image acceptance and expression. Attention to the image that the surface makes is extended by a tectonic idea of "construction of depth" that is beyond the wrapping of an outward look. Depth creates a deep area on the surface, and the surface redefines the relationship between a new sensory experience and daily life, both of which are necessary for the age.

Hapticity, Beginning of Depth
The architecture created by UnSangDong is constantly manipulating the architectural experience happening on the surface by fully utilizing the visual sensory system. In this architecture, the 2D surface is used to create an illusion and expression of the 3D world. Sometimes our eyes search a 2D surface to experience a 3D image. The materials, patterns, scales, and qualities of depths on the surface are experienced deeply by the more haptic sense beyond the eyes.

　　　A. Riegl, a prominent art historian in Germany, described his vision as "haptic (l'haptique), or made to look closer and be groped by the gaze that appeared in Egyptian relief, distinguishing it from "optical (l'optique)." The surface of this relief tightly combines the eye and hand. In addition, the flat surface allows the eye to move like a tactile sensation, and gives the ability to touch with the gaze.[1] The tactile sense is a multi-sensory experience of the building's outward look, and changes its nature. The tactile properties of the modernized surface, "touching eyes" where the tactile and visual sense meets, are closely related to the creation and surface treatment techniques of building surfaces in response to the stimulation of the senses, a consistent feature of the buildings by UnSangDong. The patterns, including the curved strip design of the Lie Sangbong Tower, perforated fold of Crystal Mountain, distorted shape seen in patterned structural mesh, cut circle of Kring, and diagonal grid of futurism, all use elements such as materials, colors, properties, textures, and patterns in a tactile manner. The implementation of this sensory surface appears in a shape where the outer element is chopped by the human scale. Pattern repetition and change show how their architecture meets the city in a way of abstract sense implementation.

도시와 만나는 방식을 보여준다. 표면은 촉각적 체험의 무대로서 도시경관을 만들어 낸다.

감각의 단층

패턴이 가지고 있는 촉각적 자극의 힘은 감각이 만나는 표면의 깊이감으로 인해 강화된다. 깊이는 도시와의 경계에서 입체적인 지각적 체험을 만드는 생성인자가 된다. 캔버스와 같이 표면에서 보여지는 즉각적인 자극, 충격, 놀람, 순간적 몰입의 희열 등의 스펙터클한2 감각의 표현은 깊이라는 건축의 독특한 특성에 의해 고정된다.

단면은 표면에 깊이를 부여한다. 중력에 저항하는 지점으로 고유한 지식 형태를 기술적으로 제공한다. 단면은 직접적으로 볼 수 없는 부분을 시각화 한다. 따라서 표면을 통해 건축을 이해하는 방법과는 다르게 기술이 포함된 추상화된 상태로 남는다. 입면과 다르게 단면은 내부와 외부를 분리하는 두께가 있는 공간이다. 은 막으로 인식된 외부의 이미지에 표면과 벽을 동시에 표시한다. 일반적으로 볼 수 없는 현상의 '잠재적 가능성'을 제공한다. 운생동이 선보인 건축의 표피가 모호하고, 다원적이고, 수사학적인 표현 또는 추상적 경관으로 특징 지어 진다면, 단면은 원칙, 엄격함을 바탕으로한 중력, 구조, 접합 등의 기술적인 언어 문법과 대한 이해로부터 가능하다. 단면은 구현의 논리를 대변하고, 구축의 기술을 내포한다. 때론 감각의 깊이를 조정하는 내부와 외부의 중재자가 된다.

운생동은 표면에서 발생하는 감각에 현대의 풍경을 만드는 그들만의 단면의 개념을 내포하게 된다. 단면의 깊이감은 유희와 가벼움을 중심으로 표현된 표면과 다르게 구축의 기술을 통해 감각적인 층위를 부여한다. 깊이는 미묘하고 역동적인 효과를 보여주고 경험자의 시각적 각도와 시간의 변화에 의해 다른 모습으로 비쳐진다. 그들의 건축에서 현상의 구축을 위해서 깊이 있는 표면을 구현하는 행위는 이미지를 실제의 경험이 느껴지는 장소로 만드는 일이다.

Gallary Yeh는 도시의 캔버스이다. 스킨 스케이프라는 개념의 단면의 틈을 의도적으로 강조함으로써 갤러리라는 프로그램의 이미지를 표면으로 표현하였다. 위에서 아래로 지지되는 것처럼 보이는 스킨은 반구조적(Atectonic) 특성을 만든다. 각기 다른 스케일로 분할된 수평의 띠와 자유롭게 절곡된 수직의 폴딩면은 자유로운 깊이감을 만들며 폴딩된 표피를 보다 가볍게 느껴지게 한다. 두께와 깊이를 통한 프로그램의 활성화는 수직적으로 갈라진 깊이에 의미를 부여한다.

Kring의 표면은 쇼룸이라는 프로그램의 특징상 내향적 성격보다는 외향적인 드러냄을 통해 그 특성과 가치를 드러내는 경우이다. 따라서 도시경관의 차원에서 홍보를 위한 강력한 아이콘으로 이미지를 만든다. 원형의 패턴은 각 다른 크기의

The surface creates an urban landscape as a stage of tactile experience.

Section of Sense

The strength of the tactile stimulus that the pattern has is enhanced by the sense of depth of the surface where it meets the senses. Depth becomes a factor that creates a stereoscopic perceptual experience at the boundary with the city. The expression of spectacle-like senses such as instant stimulus, shock, surprise, and momentary immersion, as seen on canvas, is fixed by the depth, a unique characteristic of architecture. 2)

The cross-section imparts depth to the surface. It technically provides a unique form of knowledge as a point to resist gravity. The section visualizes the parts that are not directly visible. Thus, unlike the way of understanding architecture through the surface, it remains in an abstracted state through technology. Unlike elevation, the cross-section is a space with a thickness that separates the inside and outside. It displays the surface and the wall simultaneously on the external image recognized as a thin film. It provides a "potential possibility" of a phenomenon that is not normally seen. The outward look of the architecture built by UnSangDong is characterized by ambiguous, pluralistic, rhetorical expressions or abstract landscapes. The section can be derived from the understanding of technical language grammar such as gravity, structure, and bonding, based on principles and a certain strictness. The section represents the logic of implementation and implies a construction technique. Sometimes it serves as a mediator between the internal and external so that it adjusts the depth of sensation.

The concept of section that creates the modern landscape of UnSangDong is added in the sense made by the surface. The depth of the cross-section offers a sensuous layer through the construction techniques different from those of the surface, which is expressed mainly in pleasure and lightness. Depth shows subtle and dynamic effects and is reflected by the visual angle of the person experiencing the changes of time. The architecture builds an in-depth surface upon which to construct a phenomenon, making the image a place where real experiences are felt.

Gallery Yeh is a canvas of the city. By intentionally emphasizing a gap in the cross-section of the skinscape concept, the image of the gallery is expressed as a surface. The skin that appears to be supported from top to bottom creates an "atectonic" feature. Horizontal bands divided by different scales and freely-bent, vertically-folding surfaces create a free sense of depth and make the folded skin look lighter. Activation of the program through thickness and depth gives meaning to the vertically cracked depth.

Kring's surface reveals its characteristics and value as a showroom through outward manifestations rather than introverted features. Therefore, it creates images of powerful icons for publicity in the dimensions of the cityscape. The circular pattern pushes the circles of different sizes backwards to simultaneously construct the depth and

원형을 뒤로 밀어 깊이와 패턴을 동시에 구축하나, 원형이
파동처럼 중첩되며 충돌하는 표면은 또 다시 점층적으로 뒤로
밀려난 원형으로 깊이를 만든다. 단면 디테일에서 보여지는 원형의
흐름은 기능을 담지 않는 장식적 구축의 형태로 남는다. 최근작인
몽유도원도는 외부를 수직방향으로 가로지르는 표피으로 Gallary
Yeh와의 연속선상에서 깊이감에 대해 볼 수 있다. 단지
몽유도원도는 깊이 자체를 장식적이 요소 이외에 루버라는 성능적
측면을 내포한 복합적 결과로 볼 수 있다. 단면의 형상에 장식적
깊이 덧 데어 이미지로 만들어진 공간과 연관없이 부착된
스킨이다. 보다 깊이 있는 표면은 Structural Mesh Wall에서
보인다. 대각선 패턴의 형태는 구조적인 부재와의 결합을 통해
보다 적극적인 의미의 3차원적인 시각적 경험을 가능케 한다. 선형
패턴을 만든 후 건물의 외피에 입힌 듯한 이미지는 비정형적인
개구부는 각각 다른 형상으로 표면이 된다. 깊이는 건물 내부로의
시각적 호기심을 자극한다. Seongdong Cultural & Welfare
Center 에서 보이는 사선형의 패턴의 표피는 움직임이라는 내향적
행위를 외적 표현 수단으로 사용한다. 패턴의 깊이는 움직임이라는
행위와 장식을 교묘히 혼합하며 저층부의 역동성과 고층부의
정연함을 대비시킨다. 이 표면의 깊이감은 감각적인 경험,
신비감과 그림자를 통한 시간/운동/이미지 개념을 표면에
구축한다. 미동전자의 사선 그리드는 깊이의 반복 변화를 가장
적극적으로 활용한 표피의 사례이다. 단 단면에서의 깊이가
프로그램이 가지는 특성과 보다 긴밀한 연관을 가지지 못하는
특성이 있다. 오히려 프로그램과 결합한 표피는 초기작에 가까운
Life & Power Press / Cheongsim Water Culture Center 에서
보다 적극적으로 표출되었다. 표면은 프로그램의 일부로 만든
깊이가 된다. 경계와 완충공간의 보이드는 흐름과 연관되며 계단에
의해 사선으로 분할되며 파열된 스킨의 이미지가 구현된다.
음각으로 구현된 깊이는 공간의 일부이며 프로그램이 된다.
무엇보다 주목할 만한 프로젝트로 판교주택은 깊이가 전체를
관통하는 주제가 된다. 표면의 확장된 경계 자체가 공간이 되고
은 표면이 구조가 되며 내부와 외부의 경계가 모호한 특이한
지점에 위치해 있다. 표피의 깊이는 프로그램의 일부로 모든 건축
어휘를 공간의 체험으로 통합한다.

깊이의 가치
운생동의 건축은 '구경꾼(spectator)'에게 특정한'보기(seeing)의
방식'을 강요한다. 그들의 외피는 직설적이고 감각적인 즐거움에
관한 것이다. 표면의 체험적 효과와 직접 관련되며, 화려하고
외향적인 수사학적 이미지 형식과도 연결되어 있다. 운생동이
만들어내는 표면은 기능적 존재(being)에서 외양(appearance)을
넘어 이미지(image)로서 최종의 형식을 거치며 구경꾼들에게
감각적 체험을 제공함으로써 도시경관을 재조직한다. 그러나

pattern, but the circle overlaps like a wave and the colliding surface again makes the depth into a circle that is gradually pushed backward. The flow of the circle shown in the section detail remains in the form of a decorative construct that does not function. The latest work, Mongyudowondo, has an outward look that runs vertically across the exterior, and a sense of depth on a continuum with Gallary Yeh can be seen. Mongyudowondo uses depth as a decorative element, and the louver offers a functional aspect. The skin is attached to the shaped of the cross-section without being related to the space created by the decorative depth. Deeper surfaces are visible on the structural mesh wall. The diagonal shape enables a three-dimensional visual experience in a more positive sense, through combination with the structural members. After the linear pattern is created, the irregular openings in the image of the building appear to have different shapes. Depth stimulates visual curiosity in the interior of the building. The diagonal patterned surface seen at the Seongdong Cultural and Welfare Center uses the intrinsic behavior of movement as a means of external expression means. The depth of the pattern deliberately blends the behavior and decorations, and contrasts with the dynamism of the lower part and orderliness of the upper. The depth of the surface builds sensory experience, mystery, time through shadow, movement, and a concept of the image on the surface. The diagonal grid of the "Midong Electronics and Telecommunication Headquarters" is an example of the surface that most actively utilizes the repeated change of depth. There is a feature in which the depth in one section does not have a closer connection with the other characteristics of the program. Rather, the surface, combined with the program, is more actively expressed in the near-early works of the Life & Power Press and Cheongshim Water Culture Center. The surface becomes the depth created as a part of the program. The void in the boundary and buffering space is related to the flow, and divided into a diagonal line by the stairs where images of broken skin are implemented. The engraved depth is a part of the space and becomes the program. A most notable project, the Pangyo houses become the subject with the depth penetrating the whole of the house. The expanded boundary of the surface itself becomes a space, the thin surface becomes the structure, and the boundary between the inside and outside is located at an unusual point. The depth of the skin is a part of the program and incorporates all architectural vocabulary into the experience of the space.

Value of Depth
The architecture of UnSangDong compels the spectator to adopt a specific way of seeing. The architectures' skins reflect direct and sensuous pleasures. They are related to the experiential effects of the surface and connected to the decorative and extroverted rhetorical image forms. The surface created by UnSangDong reorganizes the cityscape by passing through the final form as an image beyond the appearance of functional being, providing the viewer with a

건축의 표면은 단순히 겉으로 드러나는 광경이나 볼거리만을 의미하는 것이 아니라 그 이면에는 총체적인 경제. 사회구성체의 논리가 작용하고 있다. 단면은 시각으로부터 시작해 인체적, 지각적 경험이 출발하는 경계가 된다. 깊이는 주변세계와 신체경험의 감각을 이어주는 매체로서 표면의 역할을 다시 생각하게 한다. 깊이 있는 표면은 단지 이미지들의 집합이 아니라 이미지들에 의해 매개된 사회적 관계를 보여준다. 따라서 건축의 외피는 시각중심의 은 이미지들을 유포하는 산물이 되어서는 안될 것이다. 오히려 그것은, 실제적인 것이 되고, 물질적으로 번역된 건축의 일부로서 통합 되어야한다.

지금 우리에게 필요한 표면은 단순한 시각적 유희를 넘어 일상에 감각을 부여하는 외피이다. 표면의 패턴이 만드는 피상적이고 자극적인 시도를 넘어 표피의 깊이감을 통해 발현된 구축성이 필요한 시점이다. 표면에 부여된 깊이의 확장은 시각경험에 제한된 수용능력을 가진 인간에게 다감각적 경험을 부여하는 매체로서 역할이 가능하다. 건축의 일상에서 보여지는 생활 환경의 감각과 일치하고, 보다 깊이 있는 체험적 지각 작용을 통해 삶의 환경을 바꾸는, 깊이 있는 표면의 논리에 대한 고민이 내부로부터 필요하다. 운생동이 추구하는 얇은 감각을 넘어서는 깊이 있는 표면의 의미가 보다 적극적으로 표출되기를 기대한다.

sensory experience. However, the surface of the architecture is not merely a sight that is exposed to the outside, but the logic of the overall economy and social structure working behind it. The cross-section is a boundary from which the physical and perceptual experiences start. Depth allows us to rethink the role of the surface as a medium that connects the sense of physical experience to the surrounding world. The deep surface is not just a collection of images; it shows a social relationship mediated by images. Therefore, the outer skin of architecture should not become a product of spreading one-dimensional, visually-focused images. Rather, the skin must be practical and integrated as a part of the materially translated architecture.

The surface we need now is a skin that gives a sense of daily life beyond simple visual pleasure. This is a time when construction is expressed through the depth of the surface beyond the superficial and stimulating attempts made by the necessary surface pattern. The extension of the depth of surface can serve as a medium for imparting multisensory experiences to humans with a limited capacity for visual experience. It is necessary from the inside to worry about the logic of the deep surface that changes the environment of life through deeper experiential perception, in accordance with the sense of a living environment seen in the everyday life of architecture. There is hope that the meaning of the deep surface beyond the shallow sense that UnSangDong pursues will be expressed more positively.

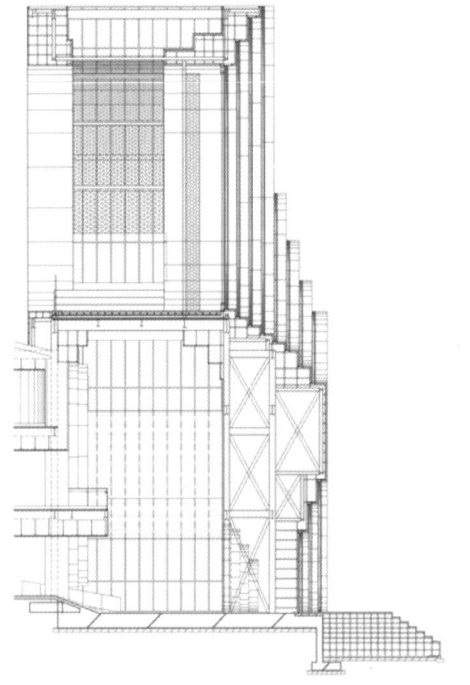

크링 단면상세도 (출처: 운생동건축사사무소)

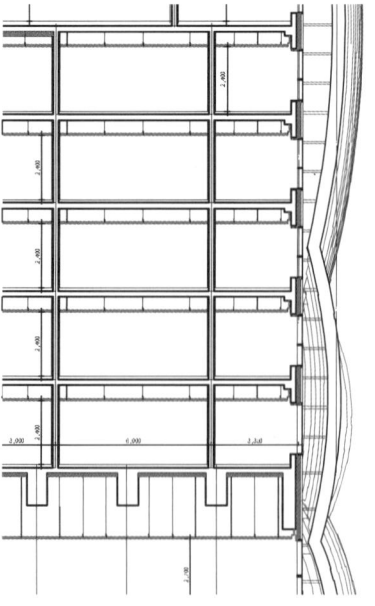

이상봉타워 단면도 (출처: 운생동건축사사무소)

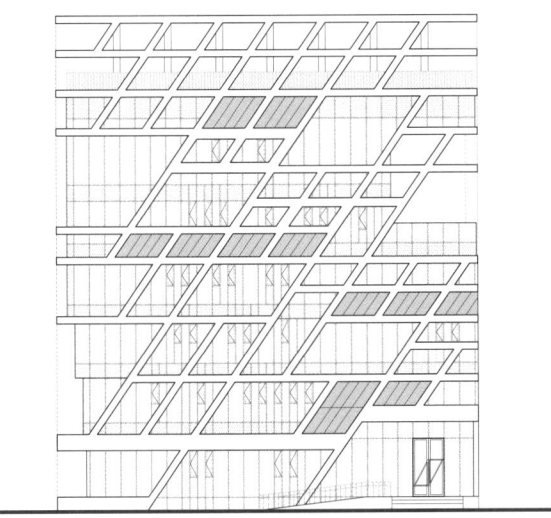

미동전자 사옥 정면도 (출처: 운생동건축사사무소)

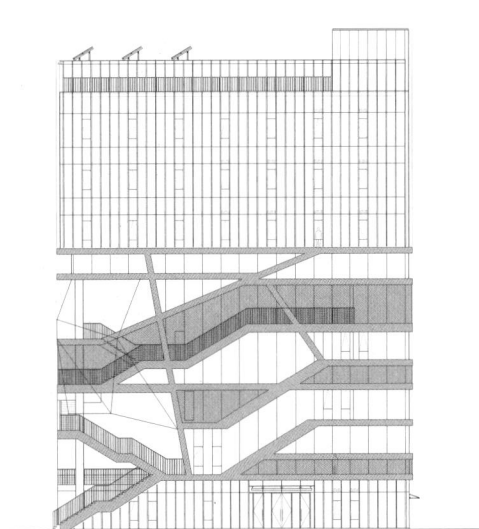

성동문화복지센터 정면도 (출처: 운생동건축사사무소)

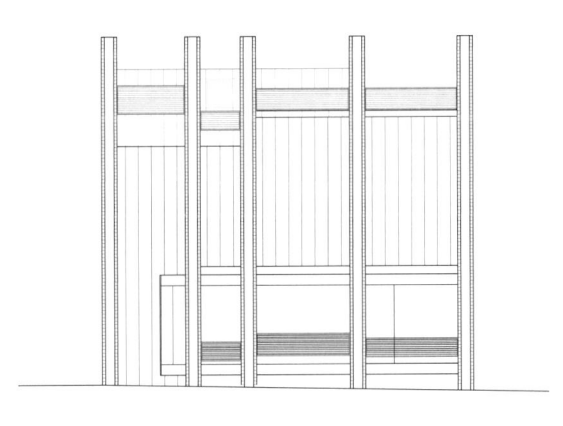

월하우스 우측면도, 단면도 (출처: 운생동건축사사무소)

1 질 들뢰즈, 하태환역, 감각의 논리, 민음사, 1995
2 스펙터클의 사전적 정의는 광경, 볼만한 것, 장관, 호화로운 구경거리, 쇼를 나타내며 간단히 호화로운(시각적)볼거리 라고 이해할 수 있다. 원래 라틴어 Spectaculum에서 유래한 이 말은 특히 프랑스에서는 연극적 재현을 지칭해온 역사를 지녔다. Barnhart,R. (1999) Dictionary of Etymology, Chambers, New York

1 Gilles Deleuze, Francis Bacon Logique de la Sensation, translated by Taehwan Ha, Minum Books, 1995
2 The dictionary definition of "spectacle" can be understood as a show or performance that is very grand and impressive. Simply, it is a dramatic (visual) attraction. Originally, it was derived from the Lain, "spectaculum." Historically it referred to theatrical reproductions, especially in France. Barnhart, R., (1999), Dictionary of Etymology, Chambers, New York.

Dasan-Dong Fortress Wall of Seoul Parking and Cultural Center

다산동 성곽길 주차장 및 문화센터

Location: Jung-gu, Seoul, Republic of Korea **Site area:** 4,275.3m² **Gross floor area:** 10,050.8m²
Building area: 2,896.9m² **Building to land ratio:** 67.8% **Floor area ratio:** 235.1% **Structure:** Reinforced Concrete
Building scope: B3 ~ 3F **Height:** 22m **Photographer:** Jaekyeong Kim(Model)

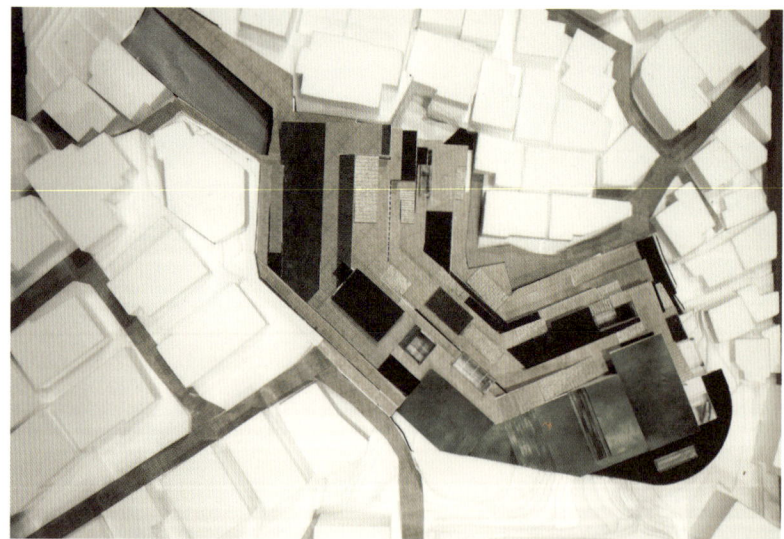

절개지형으로 단절된 도시구조와 자연환경을 복원하는 개념을 적용한다. 즉, 다산동 성곽길 주변 주택밀집지역의 절개지형으로 구성된 계획대지를 주변 도시구조와 어우러지도록 계획하기 위해서 과거의 자연지형을 복원하는 방식으로 공영주차장과 문화시설을 구성한다. 경사를 따라 줄어드는 단형의 데크 건축을 통하여 절개된 지형을 복원하여 공영주차장이면서도 조경공간이며 도시산책로의 역할을 수행하는 마을공원과 같은 공간이 되도록 구성한다.

This is a concept of rebuilding the urban structure in which the topography of the natural environment is broken up. In other words, public parking lots and cultural facilities are constructed by redesigning old natural topography in order to plan the target site composed of a densely populated area around Dasan Seonggwak-gil in harmony with the surrounding urban structures. By restoring the topography, which is broken up, through a stair-like deck construction that decreases along the slope, a space for a public parking lot, landscaping, and a town park that functions as a walkway will be constructed.

Unbuilt Compound Body: The UnSangDong Experiment

지어지지 않은 복합체 : 운생동의 실험성

Mannyoung Chung (Department of Architecture, Seoul National University of Science and Technology)
정만영(서울과기대 건축학부)

운생동의 실험성은 우리 건축에서 희소한 성향을 보여준다. 모티프는 복합체다. 2005년, 20111년, 2017년 출판된 책 3권 모두 제목이 복합체다. 정보를 배치하는 키워드와 순서도 비슷하다. 6년을 주기로 새로운 정보가 추가되고 과거의 정보를 대체한다. 거꾸로 말하면, 새로운 정보가 추가되고 과거의 정보가 대체되어도 전체 틀은 크게 변하지 않는다. 다만 끊임없이 변한다. 개별적 요소들이 아무리 다양해도, 모든 차이를 담아내면서 크게 변하지 않는/그러나 끊임없이 변하는 어떤 조직, 그것을 복합체라고 할 수 있다. 복합체는 완결될 수 없다. 모든 요소가 완벽하게 일체화되어 고정된 통합체는 더 이상 복합체가 아니다. 복합체는 불안정한 상태를 견지해야 하고, 진행 과정에 열려 있어야 한다. 질 들뢰즈의 세례를 받은 사유방식이다. 뭔가를 고정되고 정지된 것으로 사유하는 것은 벡터를 스칼라로 환원시키는 것에 불과하다는 것. 대상으로 고정되는 통합체가 아니라 끊임없이 변이하는 복합체를 지각해야 한다는 것. 이 맥락에서 분석은 명석해지기보다는 잡다해진다.

운생동이라는 명칭은 동양화 제작과 감상에서 최고의 기준으로 평가된 기운생동(氣韻生動)에서 나온 것이다. '힘과 멋이 생생하게 살아 움직이는 상태' 정도가 말뜻이다. 살아 움직이는

The experimental architecture of UnSangDong shows a rare tendency in Korean architecture. The motif is that of a compound body. All three books, published in 2005, 2011, and 2017, are entitled Compound Body. The sequence of keywords and information is similar. New information is added every six years, replacing details from the past. Yet even when new information is added and past data are replaced, the overall framework does not significantly change. It constantly shifts. No matter how diverse the individual elements are, a compound body is a structure that does not change significantly because it already contains all of the differences and is constantly shifting. The compound body cannot be completed. It stands unstable status and open throughout the process. This is what Gilles Deleuze believed. To think of something as fixed and stationary is simply to change the vector to be scalar. The compound body is not an integrator, but is ever-changing. In this context, analysis becomes more sophisticated than clear.

The name UnSangDong comes from Giunsandong, which is the best standard for making and appreciating oriental paintings. It means a state in which power and fashion come alive. What is the expression that corresponds to a compound body that is constantly changing as much as a living energy? The UnSangDong compound body is an organization with specificity, but the individuality of each

기운만큼 끊임없이 변이하는 복합체에 부합하는 표현이 어디 있겠는가? 운생동-복합체는 특이성을 갖는 조직이지만, 그 안에 여러 구성원의 개별성이 혼재되어 있다. 장윤규, 신창훈, 또 다른 누군가의 특성이 경계를 구분할 수 없는 상태에서 뒤섞이며, 변이와 움직임을 만들어 낸다. 중요한 점은 각자의 개체성이 전체 조직에 흡수되어 융해되는 것이 아니라는 점이다. 마뉴엘 데란다가 들뢰즈 해석을 통해 개념화한 조합체 이론assemblage theory은 이 점을 명확히 한다. 유기적 통일성을 이론적으로 대체하는 개념인 조합체는 "하나의 전체이기는 하지만, 그 속성은 전체를 구성하는 부분들의 상호작용에 의해 창발되는 상태"(Manuel DeLanda, A New Philosophy of Society_ Assemblage Theory and Social Complexity, 2006, Continuum, 5쪽)이다. 유기적 통일성은 내부성의 관계, 즉 "구성 부분은 전체에서 다른 부분들과 갖는 관계 자체에 의해 형성"(9쪽)되기 때문에, 전체에서 떨어져 나온 부분은 전체에 속했을 때의 부분과 전혀 다른 것이 된다. 예를 들어서 부자관계에서 아버지와 아들은 이 유기적 전체에서 벗어날 수 없으며, 벗어나면 이미 누군가의 아버지나 아들이 아니다. 반면 조립체는 외부성의 관계에 의해 규정되기 때문에, "구성 부분은 그 조립체에서 떨어져 나와, 다른 상호작용이 일어나는 다른 조립체에 끼워질 수 있다."(10쪽) 구성성분이 전체에 융해되지 않고 유지된다는 것, 전체는 창발적 결과라는 점이 좀 더 강조된다면 장윤규가 말하는 복합체는 데란다가 말하는 조합체와 같다.

　　설계사무소가 복합체라면 다양한 개별성이 부딪히며 대안을 탐색해가는 고투에서 고체처럼 굳은 동일성의 주체는 설 자리가 없다. 유연해야 더불어 나아갈 수 있다. 설계과정에서 결정적 국면은, 영향을 미치는 무수한 요인들을 담아내고, 치명적인 문제에 대응하면서, 특징이 현저한 복합체 상태를 형상화하는 순간이다. 모든 에너지가 집중되어 맞부딪치는 고투의 순간들은 거친 호흡을 수반한다. 쉽고 매끄러운 진행은 나태함의 반증일 뿐이다. 모든 문제를 잘 정리하고 세련되게 처리한 건축이 보통 완성도가 높다고 칭송된다. 하지만 역설적으로 여기서 시선은 매끄럽게 스쳐 지나간다. 통상적 감각에 순응해서 익숙해 진 건축, 스쳐 지나는 시선, 반대로 설계과정의 고투를 담고 있는 건축에서 나타나는 머뭇거린 흔적과 통상적 감각에 거슬리는 결은 스쳐 지나갈 수 없다. 생경한 건축, 긴장해서 날선 시선.

　　운생동의 복합체 이념은 실현되지 않은 2000년대 계획안들, 즉 부산타워(2003), KT&G 복합문화센터(2004), 파리 올림픽 메모리얼(2004), 광주 아시아문화전당(2005), 전곡리 선사박물관(2006), 2012 여수 엑스포 주제관(2009) 등에서 집중적으로 모색되었다. 놀라울 정도로 폭발적인 에너지를 보여준 일련의 계획안들이 풍성하게 펼쳐 보인 비전이 우리 시대의 잠재력이라면, 실현되지 않은 것은 한국 건축의 막대한 손실이다. 사회는 이미 검증되어 시야에 포섭된 것에 머무는 관성을 벗어나지 못했고, 그 수준이 고스란히 반영된 결과임을 어쩌랴.

member is mixed into it. Yoon Gyoo Jang, Cahnghoon Sin, and other people's characteristics are mixed together in a state where the boundaries cannot be distinguished, and changes and movement are created. The most important point is that individuality is not absorbed and melted down by the greater organization. Assemblage theory, conceptualized by Manuel DeLanda's interpretation of Deleuze, clarifies this point. An assemblage that is theoretically a substitute for organic unity means "being whole whose properties emerge from the interaction between parts" (Manuel DeLanda, A New Philosophy of Society: Assemblage Theory and Social Complexity, Continuum, 2006, p. 5). Because he said that "an organic unity is called relation of interiority, the component parts are constituted by the very relations they have to other parts in the whole" (p. 9). The part that comes away from the whole is completely different from the part that belongs to the whole. For example, in the relationship between a father and son, they cannot escape from the organic whole, but they are not someone's father or son, they escape from the organic whole. Conversely, because an assemblage is defined by a relationship of exteriority, "a component part of an assemblage may be detached from it and plugged into a different assemblage in which its interactions are different" (p. 10). If the component is maintained without being melted into its entity, and the whole being the product of an emergent result is emphasized, the compound body of which Yoon Gyoo Jang speaks is the same as the assemblage DeLanda described.

　　If the design office is a compound body, various individualities are encountered. In the struggle to explore alternatives, the subject of a solid identity has no place to stand. It must be flexible and go along. The decisive phase in the design process is the moment when a distinctive compound body is shaped, capturing a myriad of influencing factors and responding to fatal problems. The struggling moments where all of the energy is concentrated accompany tough breathing. Easy and smooth progress is the result of laziness. The architecture that solves all problems well is usually praised for its completeness, but paradoxically, the gaze here passes smoothly. Architecture is accustomed to adapting to an ordinary sense, and becoming a view just passing by. On the contrary, it is not possible to just pass the hesitating signs seen in architectural designs that struggle with the design process and a feeling of rebellion against ordinary sense. The result is a strange architecture and a sharp gaze with tension

　　UnsangDong's philosophy of the compound body was found in the unrealized plans of the 2000s, such as the Busan Tower Complex (2003), KTNG Culture Complex (2004), Paris Olympic Memorial (2004), Gwangju Asian Culture Complex (2005), Jeongok Prehistory Museum (2006), and 2012 Yoesu EXPO (2012). If the vision of a series of proposals showing amazingly explosive energy is the potential of our time, what has not been realized is the tremendous loss of Korean architecture. Society has already been established and has not escaped the inertia of staying within the viewpoint, and the result is where we are.

광주 아시아문화전당은 운생동의 복합체 이념을 가장 극적으로 구현한 작품이다. 계획안은 수평적인 단일 판-지형으로 구상되었으며, 128,621㎡의 도시블록을 대상으로 한 초유의 문화적 실험이었다. 대각선 길이가 570m와 360m에 달하는 대지의 양단을 들어 올린다는 발상은 연면적 173,540㎡의 기념비적 스케일에 걸 맞는 웅대한 구상이다. 수평 기준면은 중앙에서부터 넓게 경사져 오르며, 경사면에는 기존의 도시조직이 단순화되어 보이드 패턴을 형성한다. 보이드의 형상은 서로 다르고 사실상 불규칙하지만 개념적으로 반복적이어서, 패턴처럼 독해된다. 불규칙한 차이들이 거대한 조직에 흡수되면서 동일한 개념적 층위를 얻는 것이다. 도시를 안에서 바라보는 장대한 비스타가 예상되고, 도시블록의 하부구조, 조경, 광장과 길, 다양한 옥외 프로그램과 시설들, 더 나아가 경사면을 이리저리 거닐거나 머무는 사람들의 모습까지 일체화시키는 특징적인 루프 스케이프를 기대할 수 있다. 거대한 판이 깔아 담고 얹어 담은 다양한 개별적 요소들의 상호작용이 가속화될수록, 하나의 조직으로 지각되는 복합체의 잠재적 역량이 극대화되는 것이다.

건축과 지형을 수평적 복합체로 통합하는 시도는 파리 올림픽 메모리얼과 전곡리 선사박물관 등 많은 계획안들에서 확인된다. 아시아 문화전당이 차이 속에서 반복 효과를 낳는 보이드를 모티프로 사용하는 것과 대조적으로 파리 올림픽 메모리얼은 반복적인 픽셀 모듈의 수직적 변위를 모티프로 복합체 조직을 생성한 점, 두 계획안에서 사용된 경사진 지형 모티프가 전곡리 선사박물관에서는 자연적 경사 자체를 떠 있는 직선 프레임 안으로 끌어들이는 방식으로 변환된 점. 각 계획안별로 같은 주제가 다양한 방식으로 변주된 것을 알 수 있다.

부산 타워는 복합체 이념을 수직방향으로 확장한다. 마천루 계획에서 필수적인 구조 프레임을 유희적으로 변용시켜 직교체계의 경직성에 대비시켰다. 유희의 극대화는 실현의 범위를 넘어선 것처럼 보이기도 한다. 어떻게든 실현시킬 수는 있을지 몰라도 현실의 저항 또한 그리 만만치 않다.

수평-수직 복합체의 타협적 접점은 적층구조에서 탐구된다. KT&G 복합문화센터 또는 플라잉 시티는 그 전형을 보여준다. 평행 단면에 이질적 요소라 할 수 있는 극장의 경사진 볼륨을 매개로, 상업공간과 문화공간의 단면틈새에 가로, 광장, 공원을 삽입시켜 중력에 저항하는 듯 보이는 자유로운 적층을 극대화한다. 여수 엑스포 주제관에서는, 이질적인 단면상의 편차가 큰 적층 방식이, 기하학과 변형을 대비시킨 윤곽선을 통해 랜드마크적 오브제를 표현하는 방식과 결합된다.

건축 언어로 볼 때, 계획안에서 사용된 언어의 개별성은 전체 조직에 내재적인 것이 아니라 외재적인 점, 즉 하나의 계획안에 흡수, 융해되어 일체화된 것이 아니기 때문에 다른 맥락으로 얼마든지 사용될 수 있다. 문제는 계획안에서 실험된

The Gwangju Asian Culture Complex is the most dramatic implementation of the ideology of UnSangDong's compound body. The proposal was composed of a single, horizontal plate topography. It was an initial cultural experiment targeting 128,621m^2 of city blocks. The idea of lifting both ends of the land with diagonal lengths of 570m and 360m was a magnificent idea corresponding to 173,540m^2 of total floor area, a monumental scale. The horizontal base level was inclined widely from the center and the existing urban structure was simplified on the inclined plane to form a void pattern. The shape of the void was different and actually irregular, but it was conceptually repetitive so it is read like a pattern. The irregular differences were absorbed by the gigantic organization and achieved the same conceptual levels. The grand vista was expected to look inside the city and features a distinctive roofscape that unified urban block infrastructure, landscaping, squares, and streets, various outdoor programs and facilities, and even those who walked or stayed on the slopes. The acceleration of the interaction of the various individual elements laid on top of a huge plate maximized the potential capacity of the compound body, as perceived as one organization.

Attempts to integrate architecture and terrain into a horizontal compound body have been identified in many proposals, such as the Paris Olympics Memorial and Jeongok Prehistory Museum. In contrast to the Asian Culture Complex, which used a void as a motif and caused repetitive effects in the difference, the Paris Olympic Memorial created a compound body with the vertical displacement of a repetitive pixel module as a motif. The tilted terrain motifs used in the two proposals were transformed into a way of dragging the natural slope itself into a floating straight frame in the Jeongok Prehistory Museum. The same themes were varied in numerous ways for each proposal.

The Busan Tower Complex vertically extended the idea of the compound body. In the skyscraper plan, the essential structural frame was playfully transformed to prepare for the rigidity of the orthogonal system. However, the maximization of amusement seemed to be beyond the scope of realization. It may have been possible to realize it somehow, but the resistance of reality was also very persistent.

The compromise contact of the horizontal-vertical compound body was also explored in the laminate structure. The KTNG Culture Complex and Flying City are two examples. The tilted volume of the theater, a heterogeneous element in the parallel section, it maximized the free stacking that seemed to resist gravity by inserting a landscape, square, and park in a gap of a section of commercial and cultural space. In the Yeosu Expo Pavilion, a stacking method with a large deviation of heterogeneous cross-sections was combined with a method of expressing a landmark object through an outline that compared geometry and deformation.

In terms of architectural language, the individuality

복합체 이념이 다른 맥락에서 실현된 경우, 크게 위축되고, 왜곡되어 국면이 달라진다는 점이다. 예를 들어서 파리 오페라 하우스에서는 매트구조를 3차원적으로 변화시키는 간단한 수법을 사용하지만, 인터랙티브 맵이 조성하는 굴곡지형의 초점은 여전히 지형-판에 있다. 하지만 이 계획안과 같은 맥락에서 실현된 파주 생능출판사(2005)에서 초점은 지형-판이 아니라, 외피를 감고 지나는 계단 모티프로 축소된다. 공간적으로 조직 전체를 횡단하는 스케일이 거의 표면에만 적용되는 단편적 모티프로 전환되는 것이다. 같은 맥락에서 아시문화전당의 수평 무대가 서울시립대 캠퍼스 콤플렉스의 옥외 광장을 연결시킬 수 있다면, 실천의 국면에서 이념이 위축되는 또 다른 사례라고 할 수 있다. KT&G 복합문화센터의 연장선상에 있다고 할 수 있는 성수 복합문화복지센터(2005)나 (물론 시기적으로 앞서지만) 여수 엑스포 주제관과 마찬가지로 입체적 오브제의 성격이 강한 크링(2006)과 여수 엑스포 현대자동차 그룹관(2012) 정도가 이런 평가에서 벗어난다. 어쩔 수 없이 타협할 수밖에 없는 현실이 실험이 닿을 수 있는 한계라면 그것은 운생동의 한계일 수도 있다. 사실 운생동의 대표작을 계획안으로 꼽는 것 자체가 불편한 일이다.

그 불편함은 도전적 계획안과 타협적인 실현작 사이의 간극에만 국한된 것이 아니라, 2000년대와 2010년대 사이의 간극으로 인해 증폭된다. 2010년대 작품들에서는 이전에 품어져 나왔던 도전적이고 거친 결이 숨죽고 온순해진다. 파주 유치원(2014), 판교 하우스 원(2016), 한내 지혜의 숲(2017) 등 일련의 최신작은 감각적으로 쾌적하고 만족스러운 좋은 건물이지만, 폭발적인 실험적 동력과는 거리가 있다. 재능이야 여전히 돋보이지만, 완성도 높은 익숙한 언어로 위축된 것 아닌가 의구심이 든다. 2017년 현상설계에 연이어 당선한 50플러스창업지원센터와 창작연극지원센터는 여느 현상설계안과 다르지 않은 표현을 보여준다. 공공의 성격이 강화된 현상설계에서 경쟁력을 얻느라 실험성을 포기한 것인가? 사회 통념에 거슬러 불온함을 존재근거로 내세웠던 것은 한 때의 충동이었나? 실험은 경험한 적이 없는 새로운 것을 시도하는 것이고, 원만함은 경험에서 비롯한 것이다. 원만한 실험이란 형용모순에 불과하다. 원만함은 실험의 야성을 길들이고, 어디로 튀어나올지 모르는 잠재력을 통제하는 작용이다. 감각적 만족도를 느끼며 편안한 마음으로 바라보는 건축에서 실험성은 지워진다. 이 아포리아적 상황을 어떻게 벗어날 것인가 운생동의 미래에 묻는다.

of the expression used in the proposals has not been made implicit throughout the whole organization, but rather is exogenous. In other words, it has not been absorbed and merged into one plan, so it can be used in any other context. The problem is that if the idea of the compound body in the proposals was realized in a different context, it was greatly daunted, distorted, and changed. For example, the Paris Opera House used a simple technique to three-dimensionally change the matte structure, but the focus of the curved terrain created by the interactive map was still in the terrain-plate. However, in the Paju Life and Power Press (2005), which is realized in the same vein as the proposal, the focus is reduced to a staircase motif that wraps around the outer skin and is not a topography. The scale that spatially crosses the entire organization is transformed into a piecemeal motif that is applied only to the surface. In the same context, if the horizontal stage of the Asian Culture Complex could be connected to the outdoor plaza of the Campus Complex at the University of Seoul, it could be said that the ideology was intimidated in practice. However, the Seongsu Culture Complex (2005), which is considered to be an extension of the KTNG Culture Complex (even though it is ahead of schedule), Yeosu Expo, Kring (2006), which has a strong character as a three-dimensional object, and Yeosu Expo Hyundai Motor Group Pavillion (2012) are all excluded. If the reality that cannot help compromising is inevitably the limit that the experiment can reach, it may be the limit for UnSangDong. In fact, it is uncomfortable to consider the representative works of UnSangDong as proposals.

The discomfort is not limited to the gap between challenging proposals and compromised realizations, but it is amplified by the gap between the 2000s and 2010s. In the works of the 2010s, the challenging and rough texture that had been born before became calm and gentle. The latest works, such as White Cube Matrix: Paju Kindergarten (2014), House One (Pangyo House) (2016), and the Hannae Forest of Wisdom (2017) are sensationally pleasant and satisfactory buildings, but they are far from explosively experimental. UnSangDong's talent is still distinctive, but I doubt they have been intimidated by a highly accustomed language. The Generation-Convergence Start-Up Center and 50 Plus Campus and Creative Art Support Facility, which were successively elected for design competitions (2017), show similar expressions from the usual design plans. Has experimentation been abandoned in order to gain competitiveness in design competitions that strengthen its public nature? Was it a once-in-a-lifetime impulse to present improperness against social norms? Experiments involve trying new things that have never been experienced before, and amicableness comes from practice. A smooth experiment is just a contradiction. The amicableness tames the wildness of the experiment and controls the potentiality of not knowing where to bounce. Experiences are gone in architecture where people feel sensual satisfaction and look with a relaxed mind. This is the question I ask regarding the future of UnSanDong: how will this aporia be escaped?

Gallery Vogoze

갤러리 보고재
(삼성동 근린생활시설)

Location: 65-9 Samsung-dong, Gangnam-gu, Seoul, Republic of Korea **Site Area:** 746 m² **Building Area:** 371.89 m² **Total Floor Area:** 3436.62 m² **Client:** Suwon Hong Gallery Vogoze **Structure:** RC Structure **Architects:** Jang Yoon Gyoo, Shin Chang Hoon **Designer:** Kim Kyung Tae, Park Seong Yeon, Choi Yeong Eun, Kang Seung Hyun, Ko Eun JIn, Chung Chul Min, Sim Jehyun **Photographer:** Jaekyeong Kim(Model), Sergio Pirrone

보고재는 근생건축에서는 하기 힘든 조각적 건축의 의미를 결합하려고 시도하였다. 운생동의 근간작업의 화려한 재료와 조각적 변위를 우아하고 미묘한 변위로 변화시켜 그 가능성을 더욱 탐구한작업으로, 몽골리안 블랙의 화강석 사각돌을 미묘한 각도로 잘라내거나 파내어서 기존의 내부기능의 유연성과 심플함을 해치지 않은 효율성에 아트적 작업을 결합하여 기존 근생의 일반적인 이미지를 쇄하는 특별함을 실현하려한다. 더욱이 홍수원 쥬얼리디자이너가 건축주라는 점은 건축의 의도와 생활하고 작업하는 사용자의 의도가 동일시되는 시너지가 생겨난다. 보고재와 바이크 리페어 샵이 도시의 새로운 콘텍스트를 만들어내는 한 예로 작용하기를 바란다. 즉, 콘텍스트를 발현하는 새로운 방식을 시도해 보는 것이다. 본래 콘텍스트를 고려한 건축이란 특정한 건축물에 관계하는 역사적, 문화적, 지리적인 배경이 되는 조건 등을 고려한 건축 만들기를 하는 것이라 볼 수 있다. 그렇지만 현대의 콘텍스트는 주변 건축과 유사한 건축으로 유지시키는 것이 아니라 사회적인 역할과 변화의 주축이되는 건축적 이슈를 생성하는 것이 될 필요가 있다. 건축이 완성되는 순간이 주변과 관계상 낯설음으로 시작되지만 이것의 영향으로 도시가 변화되는 새로운 촉매제로서 작용하는 시발점이 될 수 있다.

VOGOZE includes the meaning of sculptural architecture, which is difficult to apply to a neighborhood building. We further explored the possibilities by transforming colorful materials and sculptural dislocation into elegant and subtle displacement. Mongolian Black Granite stones are cut out or dug into subtle angles to offset the general image of the existing neighborhood by combining artistic work with efficiency that does not interfere with the flexibility and simplicity of the existing internal functions. Moreover, the fact that jewelry designer, Soowon Hong, is the owner of the building boosts the synergy and the intentions of the architecture and the user, who uses and works in this building, are equalized. It is hoped that Gallery VOGOZE and a bike repair shop will serve as examples of creating a new concept for the city. In other words, it is a new way of expressing the context. The architecture that takes into consideration the context considers the conditions that become history, and cultural and geographical backgrounds of a specific building. However, contemporary contexts need to create architectural issues that are the mainstay of social role and change rather than erecting buildings similar to those around them. The moment when the construction is completed begins the unfamiliarity with the surroundings, but it can be a starting point for a new catalyst to change the city through the influence of unfamiliarity. Various architectural issues are settled in the city and become a driving force to make the city more alive and meaningful.

Midong Electronics & Telecommunications Headquarter Office

미동전자 사옥

Location: Yeoksam-dong, Gangnam-gu, Seoul, Republic of Korea **Site area:** 949.9 m² **Program:** Office, Shop **Building area:** 473.55 m² **Building to land ratio:** 49.80% **Gross floor area:** 4,022.71 m² **Gross floor ratio:** 285.70 % **Height:** 24.50m **Architect:** Unsangdong Architects **Design Team:** Sangho Jeong, Samyeol Yoo, Seunghyun Kang, Soohoon Choi, Minkyun Kim **Design period:** 2014.12~2015.04 **Completion date:** 2016.04 **Photographer:** Unsangdong(Model), Sergio Pirrone

사옥이란, 기업의 문화를 이해하고 구성원의 행복을 담는 건축이며 기업의 브랜드를 담고 미래의 가치와 꿈을 실현하는 건축이다. 그러므로 기업사옥은 소비자 및 시민과 소통하는 가장 중요한 기업정신을 담아야 한다.
끊임없는 변화와 혁신을 통해 지속적으로 성장하는 미동전자의 기업이념을 담았다. 건축 또한 기존 건축의 형태를 답습하지 않으며 미래의 비전과 혁신을 담는 건축을 제안한다.
또, 무겁고 권위적인 사옥이 아니라 구성원의 밝은 행복을 담는 건축을 추구했으며, 친환경적 건축 공간 및 형태를 제안하며, 내부적으로 밝고 따뜻한 건축공간을 제안했다.

A headquarter office is the place that understands the culture of the company and delivers comfort to workers; an architecture that embodies the corporate and realizes future value and dream. It should embody the essence of the corporate spirit that communicates with consumers and citizens.
We aim to embody the corporate philosophy of Midong Electronics - to improve through continuous change and innovation. The building does not follow the form of existing architecture, rather embodies innovation. And we pursue an architecture that carries happiness to workers rather than a heavy, authoritative building. We propose eco-friendly space and warm, bright interior.

Rare Architecture
정의하기 어려운 건축

Hangman Zo (DAAE, Seoul National University/ TAAL Architects)
조항만 (서울대학교 건축학과/ 탈건축)

정의하기 어려운 건축[1]

장윤규와 신창훈이 이끄는 운생동UnSangDong은 그들이 지난 십칠 년 간 발표해온 파격의 다양한 건축물처럼 한 문장으로 정의하기 어려운 건축집단이다. 작품들을 통해 계속 변화하는 그들의 작가적 성향 속에서 '운생동은 이러하다'라는 도시와 건축, 인간과 환경에 대한 일관된 그들만의 생각을 짚어내는 것도 쉽지 않다. 또한 그들이 생산해내고 있는 건축물들에 대한 작가 자신의 설명이나 수많은 전문 비평들도 작품을 보는 틀이나 강조하는 점이 작품마다 다르다. 사무실을 개설한 이후 지금껏 130여 개의 프로젝트를 완성했다고 하니 그 다양한 종류만큼이나 프로젝트의 수도 많다. 이 수많은 다양한 작품 중 어떤 것들을 선택하고, 비평하는 것은 좁은 지면에서 쉽지 않다. 건축사建築史라는 조금 큰 틀과 긴 시간 속에서 건축집단 운생동의 탄생과 의미, 작업 태도, 그리고 2018년 지금 그들의 위치를 어슴푸레 가늠해보는 것이 그나마 가능하고 의미있는 일 일 것이다.

시공과 사건이 혼존하는 상변화의 시대

138억년 전 특이점Singularity에서 대폭발Big Bang으로 시작된 우주는 지금도 팽창하고 있다. 우주가 다시 수렴하여 빅크런치Big

Rare Architecture[1]

UnSangDong, led by Yoon Gyoo Jang and Changhoon Shin, is an architectural group that is difficult to define clearly in one sentence like a series of unconventional buildings that they have produced over the last 17 years. It is not easy to discover their consistent thoughts about the city, architecture, humanity, and the environment, and be able to say "UnSangDong is like this" because of the architects' tendency to change continuously through many works. In addition, the architects' own explanations of the buildings they create and the numerous professional reviews differ from one work to another. Since opening the office, they have completed about 130 projects so far, and there are as many projects as there are kinds of projects choosing and criticizing any of these many and varied works is not easy. In the long architectural history, it will be meaningful and possible to measure the birth, meaning, attitude of the group and the position of "UnSangDong" now in 2018.

The Era of Phase Change where Time, Space and Event are coexisting

The universe started with the big bang from a singularity about 13 billion years ago, is still expanding. Nobody knows if the universe will converge and begin again with a Big Crunch, face the end with a Big Freeze, or a Big Rip as it

Crunch를 거쳐 빅뱅으로 새로 시작될 지, 계속 팽창하여 빅프리즈Big Freeze나 빅립Big Rip의 끝을 맞을 지 누구도 알 수 없다. 안다고 해도 아주 먼 미래의 일이라 확인할 인류는 없다. 우주의 미래는 불가지의 영역이니 차치하더라도, 광대한 은하의 티끌 같은 45억 년 된 창백한 푸른 점 The Pale Blue Dot[2] 위, 35만 년 된 사피엔스의 삶이나 7천 년 된 인류 문명의 역사가 앞으로 어떻게 변해갈 지 누구도 장담할 수 없다. 30여 년 전 프랜시스 후쿠야마 Francis Fukuyama는 '역사의 종말'에서 이데올로기는 자유민주주의 Liberal Democracy로 종점에 도달하고 이것이 인류 마지막 정부 형태일 것이며 그로 인해 역사의 진보는 종말을 고할 것이라고 호기롭게 주장하였다. 그렇다면 그가 말한 역사 종말의 시대를 통과하고 있는 최후의 인간[3]은 아마도 현시대, 지금을 살아가는 우리들일 것이다. 한 단계 한 단계 거치며 상당한 인과관계를 가지고 점진적으로 진보해왔던 건축을 포함한 인류 문명의 역사는 21세기 들어 폭발적인 변모를 보이고 있다. 상상을 뛰어넘는 과학기술, 그리고 디지털로 이루어진 또 다른 세계를 만들어 낸 최후의 인간들은 지금 과거와 현재를 공존시켜 시간을 무력화하고, 공간과 거리를 뛰어넘으며, 모든 사건과 행위를 기록하여 박제한다. 생명은 4가지 단백질 조합으로 된 기호의 지도로 그려냈고, 모든 살아있는 유기체 중 최초로 무기체와 결합을 시작하였다. 또한 인간도 계속 변하여 이제 신이 되고자 한다는 유발 하라리Yuval Noah Harari의 주장[4]을 접하지 못한 사람이라도 인류가 이제 어떤 상변화의 문턱을 넘고 있으며 다음에 대략 어떤 것이 오리라 추측할 수 있었던 관성적 진보의 시대는 완전히 저물었다고 느낀다. 인간, 역사, 도시, 지구의 자연 환경 등 모든 맥락이 동시에 혼존하여 맥락들이 사라지는 시대로 변모하는 지금, 이 같은 극적인 변화 언저리에서 운생동은 그들의 도전적이고 예측 불가능한 작업을 시작한다.

맥락을 고려하지 않는 지금의 건축

이런 시대 상황에서 '자기복제식의 건축을 경계한다', '컨텍스트를 고려하지 않는다'는 장윤규의 말[5]은 가볍지 않은 무게로 시대와 공명한다. 무책임하게 들리지 않는다. 그가 작품을 통해 내세운 '새로움', '백과사전', '반응체', '상상체', '복합체', '신화', '클립시티', '경계를 넘은 통합' 등의 키워드는 모든 시간과 공간, 그리고 사건이 동시에 현존하려고 하는 '바로 지금'에 집중에서 만들어진 것이다. 오랜 시간 속에서 생성된 맥락은 운생동에겐 우선 고려 대상이 아니며, 맥락의 고의적 배제와 회피로 인한 그들 작품의 첫 인상은 '뜬금없다'라는 것이다. 판단의 중요한 근거인 맥락을 알아낼 수 없는 운생동의 작품은 대중에게 낯설 수 밖에 없다. 게다가 강렬한 조형과 새로운 컨셉으로 인해 그들의 건물은 마치 평화로운 들판에 폭탄처럼 갑작스레 내리 꽂혀 폭발한다. 이것은 충격적인

continues to expand. We will not be able to see which one it will be because it is in the very distant future. The future of the universe is unknowable. No one can know how the 350,000-year-old Homo sapiens, and the 7,000-year-old human civilization on our pale blue dot[2] like a speck of dust in the vast galaxy, has changed over the 4.5 billion years. More than three decades ago, Francis Fukuyama argued vigorously that the demise of history would reach the end with liberal democracy. This would be the last governmental form of mankind and then the progress of history would come to an end. If so, the last human being who will pass through the age of the end of history are probably the ones[3] who are living now. The history of human civilization, including architecture, that has developed step by step, has progressed with considerable causality, and is showing an explosive transformation in the 21st century. The last human beings who created another world composed of advanced science, technology beyond imagination, and digital applications are now able to neutralize time coexisting in the past and present, overcome space and distance, and record all events and actions. Life has been interpreted to a map of symbols composed of a combination of four proteins. It was the first of all living organism to begin to unite with inorganics. Even those who do not agree with Yuval Noah Harari's assertion[4] that humans are constantly changing into gods have now felt that the age of inertial progress, in which mankind overcomes a threshold of change and then assumes that something will come about, is completely over. Now, with all the contexts of humans, history, cities, and the planet's natural environment coexisting together and turning into an era in which the contexts disappear, during these dramatic changes, UnSangDong begins their challenging and unpredictable practice.

Context-less Architecture of Now
In this age, the words of Yoon Gyoo Jang "Stay away from self-replicating architecture" and "Do not consider context"[5] resonate with the times with the considerable weight. Key words such as "novelty," "encyclopedia," "reaction body," "imaginary body," "compound body," "myth," "clip city," and "integration beyond the border," which he introduced through his works, are made to focus on "now" where all the time, space, and events are about to exist at the same time. The context created over a long period of time is not the priority for UnSangDong, and the first impression of their work due to the intentional exclusion and avoidance of context is "out of the blue." The work of UnSangDong, which does not depend on context as an important basis of judgment, is unfamiliar to the public. In addition, due to intense modeling and new concepts, their buildings look like a bomb exploded in a peaceful field. This can be summed up as an attempt to create a context that did not exist because of its impact on the surroundings with shocking newness.

새로움으로 주변에 영향을 끼쳐 여태 존재하지 않던 컨텍스트를 만드는 시도로 요약될 수 있다.

신/이종의 미학

그리하여 운생동의 건물을 접할 때의 감정은 '강렬하고 낯선 새로움'이다. 마치 만나보지 못했던 미지의 생명체를 접하는 그런 감정인데 거기에는 '와 저게 뭐지?' 하는 경탄에 호기심, 두려움, 역겨움, 생경함, 때로는 친근함까지 가미되어 설명하기 힘든 인지부조화의 지점을 만든다. 현대미학의 언캐니Uncanny와도 또 다른 이런 건축적, 미학적 감흥을 '신/이종의 미학'이라 할 수 있을 것이다. 하지만 오해는 하지 말자. 그럼에도 불구하고 운생동의 건축물은 하나하나 완성도 있는 계를 이루고 있으며 그 작동에서도 대부분 성공적이다. 이것은 건축가가 인간과 삶, 건축과 도시가 만나는 지점인 '바로 지금'에의 집중이 현명한 선택이었다는 방증이다.

새롭기 위하여:
관계없는 것들의 통합을 통한 만연하는 '기본'의 재정의

의도적인 탈맥락과 더불어 운생동의 건축을 새롭고 낯설게 보이게 만드는 다른 이유는 그들의 건축이 이미 완연한 모든 '기본'들에 도전하는 답이기 때문이다. 운생동의 다양한 작품들은 그 구체적인 해법에서보다는 내재된 건축가의 질문에 의해 보편성을 획득한다. 질문은 언제나 그 답보다 보편적이기 때문이다. 사회나 정치, 인문이나 예술에서 하나의 질문에 여러 개의 답이 가능한 이유이기도 하다. 프로젝트를 질문에서 시작한다는 운생동은 도전할 기존의 보편적인 질서를 먼저 파악하고 골라낸다. 그리고 그것을 '기본'이라 부른다. 예를 들어 근대 '건축의 기본'인 그 꼬르뷔제 Le Corbusier의 도미노 시스템 Dom-ino System에 던진 운생동의 질문은 확연히 구분된 기둥과 슬라브로 된 구조가 아닌 벽만을 이용하여 기둥을 없앤 구조(크로노토프 월하우스, 2017)라는 해답을 만들어냈고, 내/외부를 가르는 얇은 경계라는 '외피의 기본'에 대한 도전은 벽과 지붕과 스킨이 구분 없이 일체화되어 에너지를 생산하고 부하를 줄이는 주름진 지붕을 가진 집(Rooftecture, kolon E+Green Home, 2012)이라는 구체적인 해답으로 제시되었다.

매번 새로운 해를 만들기 위한 운생동의 전략 중 가장 자주 보이는 것은 크게 관계 없어 보이는 것들의 통합이다. 건축물을 안전하게 서있게 하는 최적의 시스템이라는 '구조의 기본'에 대한 운생동의 해답으로 제시된 오션어스 해운대 본사, 2014에서는 회사의 아이덴티티를 구현하기 위해 과장된 패턴의 비효율적인 구조 프레임과 대양과 마주한 깊이 있는 조각적 외피와의 통합을 이용하였고, 금호그룹의 문화 컴파운드 Kring, 2008에서는 파사드와 브랜드, 건축주의 기호가 여러 개의

Aesthetics of New Heterogeneity

Therefore, when an UnSangDong building is first encountered, the feelings are "intense and strange newness." It is such a unique emotion like seeing an unknown creature. It creates a point of cognitive disharmony that adds to curiosity, fear, nausea, ecstasy, sometimes intimacy, and the wonder of "Wow, what is that?" Unlike uncanny in modern aesthetics, this architectural and aesthetic inspiration can be called the "aesthetics of new / heterogeneity." But please do not misunderstand. The architecture of UnSangDong is a complete work of art, and most of its works are also successful. This is a proof that it was a clever choice for the architects to focus on "right now," where the human, life, architecture, and city meet.

Redefining of a Pervasive 'Basic' through the Integration of Irrelevant Things for being New

Another reason for the new and unfamiliar architecture of UnSangDong, such as intentionally out of context, is to answer the challenge against all the "basics" of their architecture. The various works of UnSangdong acquire universality through the question of the architect rather than the specific solutions. The question is always more common than the solution. This is why many answers can be given to a single question in sociology, politics, humanities, and art. UnSangDong begins a project with a question, and then identifies and selects the existing universal order to challenge. And it is called "basic." For example, the UnSangDong's question about Le Corbusier's Domino System, which is the basis of modern architecture, has given the answer as a structure in which columns are removed and only walls are used (Chronotope Wall Houses, 2017) rather than making a clearly separated column and slab structure. The challenge of the "basic of the surface," which is a thin boundary between inside and outside, is presented as a concrete answer for a house with a corrugated thick slab that produces energy and reduces the load by integrating walls, roofs, and skins.

One of the most frequent strategies of UnSangDong for creating extraordinary solutions each time is the integration of things that seem to be largely irrelevant. Ocean Earth's Haeundae Headquarters (2014) was presented as a solution to the "basic structure" of an optimal system that helps buildings stand safely. An inefficient structural frame of exaggerated patterns was used along with an integration of deep and sculptural surfaces facing the ocean to imclude the company's identity. In the cultural compound of Kumho Group (Kring, 2008), the facade, the brand, and the owner's symbol were integrated into a geometric form of multiple concentric circles. This presented a new species of urban landscape and sharply asked the question "What is the basis of the landscape of the modern city."

동심원이라는 기하학적 조형으로 통합되어 도시풍경을 새로운 종을 제시하며 '현대 도시의 경관의 기본'에 대하여 날카롭게 질문하였다.

원전의 자유로움을 누리는 건축가

주지하다시피 운생동은 프로젝트마다 전작의 성취에 의존하거나 구속되지 않고 또 다른 새로움과 다양성을 계속 획득해 왔다. 이것은 자기 복제를 회피하려는 노력과 더불어 그들의 개개의 작품이 원본성을 획득하고 있기 때문이며, 원전을 쓸 수 있는 건축가만의 자신감에서 기인하는 것이다. 식민시대 베트남에서 보자르 스타일에 현지에서 발견한 무언가를 자유롭게 섞어 독특한 인도-차이나 스타일을 만든 건축가들은 에꼴 데 보자르-출신이었다. 보자르라는 도그마를 깨기에 보자르의 적자들만큼 적당한 이도 없다. 스스로 만든 성공적인 개념, 의도, 선언 등은 오로지 스스로에 의해서만 자유롭게 수정되고 부정될 수 있음을 운생동은 쌓여가는 포트폴리오를 통해 증명하고 있는 것이다.

보이는 것이 실재는 아니다[6]

운생동은 파격적 건축물들은 많은 관심과 함께 과도한 조형과 조각 같은 건축이란 비판도 받아왔다. 또한 그 점으로 많은 관심과 함께 인구에 회자되었다. 그러나 보이는 것으로만 건축을 평가할 수는 없다. 미학은 개인마다 다른 관심사이고 아무리 삐까번쩍하는 건물이라도 결국엔 현실생활 – 삶에 뒤덮일 운명이어서 건축의 실재 가치는 시간과 함께 서서히 드러난다. 운생동에겐 지금 답해야 할 매우 중요한 두 가지 질문이 주어져 있다. 첫 째는 '운생동을 만들어온 지금까지의 전략을 지속할 것인가?'에 대한 것이다. 계속 이어진 충격이 일상이 되면 더 이상 충격이 아니게 된다. 그러나 첫 질문의 답이 '예' 라면 두 번째 질문은 '그렇다면 충격의 일시성Ephemerality을 어떻게 극복할 것이냐?'가 될 것이다. 이 두 질문에 대한 운생동의 답과 그 방향에 따라 30년 뒤 한국건축사에서 도달할 운생동의 종착역은 달라질 것이다. 앞서 말한 '원전을 쓸 수 있는 건축가'는 지금 대한민국에서 손에 꼽힐 정도이다. 그 위치와 지위에 걸맞는 책임과 의무 중 가장 중요한 것은 그들이 생산하고 있는 원전이 건축사에서 진정 의미 있는 작업이 되도록 하는 것이다.

---

1 칼 세이건의 저서의 이름으로 보이져 1호가 1990. 02.14 태양계를 떠나며 보내온 사진에 감명받아 지구를 설명하며 붙인 말.
2 후쿠야마는 역사를 최초의 인간이 최후의 인간이 되어가는 과정으로 보았다.
3 그의 저서 Homo Deus 참조
4 조선비즈 인터뷰 '자기복제식 건축은 안 한다' 2013. 09. 14
5 이탈리아의 천재 물리학자 카를로 로벨리의 저서명

Architects who Enjoys Freedom of the Original

As you know, UnSangDong has continued to create newness and diversity without depending on the accomplishment of a previous work or being bound by each project. This is because their individual works express originality with an effort to avoid self-reproduction. In addition, it takes a confident architect to be the original. Architects who created a unique Indian-Chinese style by freely mixing something found locally in the Beaux-Arts style in colonial Vietnam were from ecole des Beaux-Arts. There is nothing as good as Beaux-Arts's heritor to break up the dogma called Beaux-Arts. UnSangDong is proving through its portfolios that successful concepts, intentions, and declarations created by them can be freely modified or denied solely on their own merits.

Reality is not what it seems[6]

UnSangDong has been criticized not only for its interest in preposterous buildings but also for its excessive form and sculpture. It also created much interest among people. However, architecture cannot be evaluated only by what is seen. Aesthetics is a matter of individual concern, and even a brilliant building is destined to be covered by real life in the end, so the actual value of architecture gradually evolves with time. UnSangDong has two very important questions to answer now. The first one is, "Will you continue the strategy you have made so far in UnSangDong?" When the shock becomes a routine, it will no longer be a shock. However, if the answer to the first question is yes, then the second question will be, "How does one overcome the ephemerality of shock?" Depending on the answer to these two questions and the direction taken by UnSangDong, the ending destination of UnSangDong, which will be reached in Korean architectural history after 30 years, will be different. The aforementioned "architect who can use the original" is rarely present in Korea now. One of the most important responsibilities and duties of their position and status is to make sure that the original they produce are truly meaningful architectural works.

---

1 The architecture of UnSangDong is divided into two types. This is a public project that links the modern architectural traditions with winning projects of different characters and sizes in design competitions in collaboration with architectural design companies, as well as outstanding private projects with self-defined experimental concepts such as compound body and imaginary body. This article focuses on the first type.
2 When Voyager 1, left the solar system on February 14, 1990, Carl Sagan was impressed with the pictures sent back that explained the earth.
3 Fukuyama saw history as a process by which the first human becomes the last human.
4 Refer to his book Homo Deus.
5 Chosun Biz interview "Do not build self-replicating architecture." 2013. 09. 14
6 The Book of Italian Genius Physicist Carlo Lobelli

White Quarter Circle
청심유치원

Location: 771-6, Yeoksam-dong, Gangnam-gu, Seoul, Republic of Korea **Use:** Educational Facilities, Commercial **Site area:** 577.50m² **Building area:** 322.48m² **Gross floor area:** 2,095.79m² **Building scope:** B3 ~ 5F **Height:** 26.61m **Building to land ratio:** 55.84% **Floor area ratio:** 199.83% **Structure:** Reinforced concrete structure **Architects:** Unsangdong Architects (Yoongyoo Jang, Changhoon Shin) **Design team:** Gyeongtae Kim, Soohoon Choi, , Jeongseop Kim **Interior team:** USD Design Group (Jaehyeon Shim, Byeongu Kim, Eunyeong Hwang) **Photographer:** Jaekyeong Kim(Model), Yongkwan Kim, Jaeyoun Kim

대지의 물리적 조건을 보면 도로와 면한부분이 15m, 깊이가 38m로 단변대 장변의 비율이 1: 2.53으로 좁고 긴 대지위에 계획되어진다. 도심속 어린이학교는 어린 학생들을 위한 배려의 건축이며 창의적인 상상력이 작동하는 건축을 제안한다. 꿈을 키워가고 세상을 배워가는 곳으로 이곳은 사회를 경험하는 첫 공공적 만남의 장소이다. 건축은 친근함과 생동감이 결합된 디자인속성을 모티브로 어린이 감성을 담는 입체적 플랫폼 학교를 제안한다. 학교를 다니는 어린이들과 인트렉티브한 감성의 교감체로 반응하는 건축이길 제안하였다. 어린이들의 밝은 동심과 상상력이 건축에 투영되어 다양한 별명으로 은유되길 바랬다. 도심속에 '1/4 circle' 기하학을 디자인 모티브로 삼는다. 이는 circle의 분절, 유희적 변형과 구성을 통해 전혀 다른 매스와 입면을 만들어 낸다. 보는 사람들의 감성에 의해 자연의 새싹처럼 불리기도, 꿈의 성처럼 읽히기도, 밝은 에너지가 투영되고 인트렉티브한 감성이 소통하는 창의적인 장소성의 건축을 만들려 했다.

The physical dimensions of the target site are 15m long and 38m deep, facing the road. The ratio of the short side to the long side is 1: 2.53, which is a narrow and long piece of land. The urban children's school is a building for caring for young students and should be a building where creative imagination works. The school is the first public where children experience society, dream of the future, and learn how to live. The architect proposes a three-dimensional platform school that embraces children's emotions with a motif of design attributes that combine familiarity and liveliness. We hope that children will interact with the building. Also, we hoped that the bright innocence of childhood and the students' imaginations would be projected into the architecture of the building and be delineated by various nicknames for the building. The "1/4 circle" geometry is used as a design motif in the city. This creates a completely different mass and facade through the circle segment, a playful transformation, and composition. Depending on how people feel about the building, we tried to make architecture a creative space where it can be called a sprout of nature or a dreamy castle. In other words, we tried to create a space where bright energy is projected and numerous feelings interact.

UGANDA Healing Mountain

우간다 힐링 마운틴(Healing Mountain) 청소년센터

Photographer: Jaekyeong Kim(Model), Unsangdong

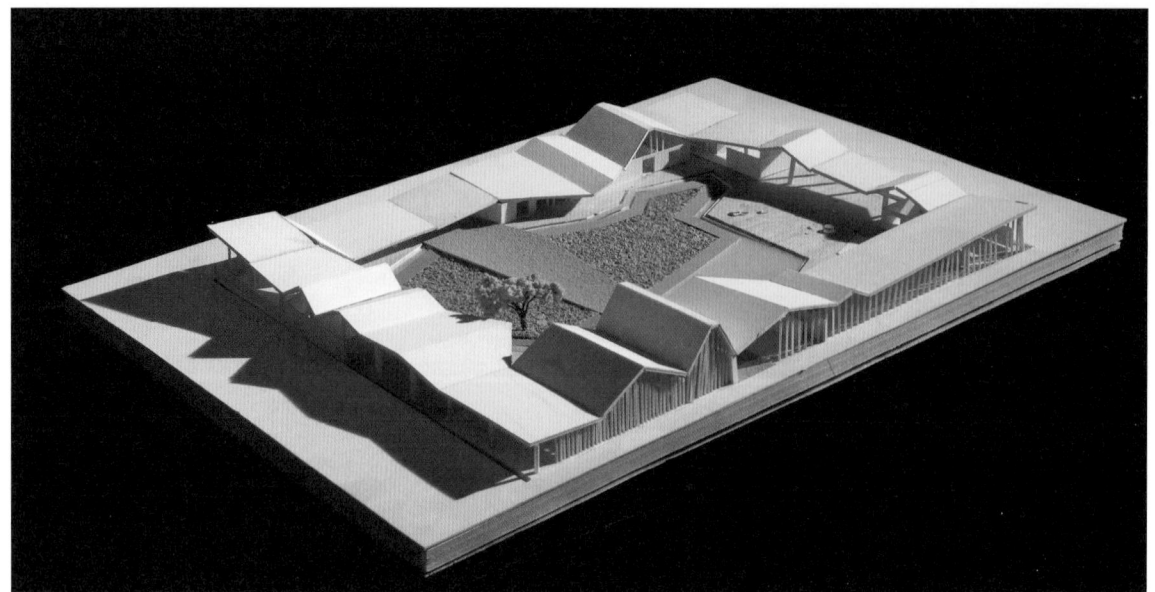

아프리카 우간다의 중북부에 있는 작은 마을 아무리아라는 작은 마을에 에이즈 감염 아동 위한 시설을 구축한다. '우간다 힐링 마운틴(Healing Mountain) 청소년센터'는 정하희 선교사님의 헌신과 한국 국제기아대책기구 그리고 임종범씨 지원으로 지어진 건축으로, 에이즈 감염 아동을 위한 종합복지센터다.
아프리카에서도 오지에 속하는 이곳은 건축이란 인공이 존재하지 않는 원시의 땅이다. 전기는 물론이고 식수도 빗물 받아 겨우 해결한다. 원시적 로우테크를 이용한 건축을 구축한다. 현대적 기술이 아닌 인간의 노동을 통해서 획득되는 공간이며 가장 원초적인 건축을 구현하려 하였다. 그 땅에서의 흙으로 벽돌을 만들어내고 그땅에서의 나무를 엮어 지붕을 만들기 위한 트러스를 구성하였다. 건축교육도 현장경험도 없는 현지인들의 땀과 노력으로 이 건축은 완성되었다. 아프리카지역에서 깨끗한 물을 얻기 위한 골을 구성하기위하여, 다양한 변위를 가진 박공지붕의 형상으로 연속하는 지붕으로서의 건축을 제안한다. 끝없이 펼쳐진 초원에 아이들에게 꿈과 희망을 주는 작은 산들의 집합으로 구성하였다. '힐링 마운틴(치유의 산)'이란 이름은 산 모양 지붕을 의미하는 동시에 어린 아동들을 위한 치유의 공간이 됐으면 하는 바람을 담고 있다. 중정형의 평면구성을 통하여 아늑하고 안전한 활동과 교육을 할 수 있는 공간을 구성한다. 중정형 평면에는 치료실, 교육실, 사무실, 교회 등으로 이루어져 있으며 실기능에 따라 높이가 다른 박공형 지붕을 구성하였다. 대지 1만3643㎡(약 4127평), 연면적 1584㎡(약 479평)의 단층으로 구성되며, 냉방시설이 없는 점을 감안해 지붕과 벽면을 1m 정도 띄워 통풍이 잘되게 했다. 기계 도움 없이 벽돌 한 장씩 나르고 망치로 일일이 못질해 만들어낸, 하이테크(high-tech)가 판치는 세상에서 로테크(low-tech)로 만든 기적 같은 건축이다.

This facility was built for children infected with AIDS in Amuria, a small village in the middle north of Uganda. "Uganda Healing Mountain Youth Center" is a comprehensive welfare center for children infected with AIDS, and is a mission of Hahee Jung, Korea Food for the Hungry International, and has support from Jongbeom Im. In Africa, this place is situated in a remote area and there is no artificial architecture in this primitive land. There is no electricity and drinking water is obtained from rain water. We built a building using primitive low tech. We tried to create a space by human labor, not modern technology, and build a very raw construction. We made bricks with soil from the land and trusses to make a roof by weaving wood from the ground. This building was completed with the sweat and effort of the local people who had no architecture education or field experience. In order to construct a valley to obtain clean water, we proposed a building with a continuous gabled roof with various displacements. The architecture consisted of a collection of small mountains giving dreams and hopes to children in an endless meadow. "Healing Mountain" means a mountain-shaped roof and is also a healing area for young children. It forms a space for comfortable, safe activities and education through a flat composition courtyard. The plane of the courtyard consists of a treatment room, a training room, an office, a church, and other rooms, and has roofs with different heights depending on the function of the rooms. The building consists of a single level on 10,436m^2 of land with 1584m^2 of total floor area. Considering that there is no air-conditioning system, the roof and the wall were separated about 1m to create ventilation. It was a miraculous low-tech building in a world of high-tech where people carried the bricks one by one without any help from a machine and hammered the nails.

Hetero-Pragmatics of Architecture
이질적 실용의 건축

Helen Hejung Choi (School of Architecture, Kookmin University)
최혜정 (국민대학교 건축학과)

사실 원고를 의뢰 받고 운생동에 대해 이야기하는 것에 대해 고심이 많았다. 긴 시간 가까이서 봐 왔기 때문에 더 조심스러운 점도 있었고, '운생동의 건축을 어떻게 봐야 할까'라는 질문은 어렵기 때문이다. 그들의 건축을 가까이서 보는 동안 드문드문 이 질문을 떠올렸지만 명쾌한 답을 생각해내지 못했다. 그래서 이 원고의 시작은 내가 아직 찾지 못한 이 질문의 답을 보류한 채, 운생동의 건축에 대한 여러 가지 오해와 편견에 대해 먼저 얘기해보고 싶다. 여기서 오해라 함은 - 운생동은 건축적 요소를 착실히 배합하여 훌륭한 체계를 만들어내는 모범적 건축보다 즉물적으로 반응하거나 튀는 건축을 한다, 혹은, 이론, 철학적 사유를 담은 텍스트가 건축물을 에워싸고 있는 듯한 글과 실행 사이의 이질감이 있다 - 같은 평가를 가리킨다. 이 오해는 어느 정도 사실일지도 모른다. 생각해보면 운생동은 그들 스스로를 어떤 역사적, 담론적 맥락에 성실하게 배치하는 부류도 아닐뿐더러, 텍스트에서 인용되는 많은 레퍼런스나 개념들은 그들의 작품에 대한 고민보다 그들이 자체적으로 풀어내는 사회 태도적 고민에 가깝게 들린다. 그들이 만드는, 혹은 작업이 만들어지는 것에 대한 성실한 설명과는 그 결이 다르다는 느낌을 지울 수 없다. 운생동에 대한 이런 크고 작은 굴절들은 때로 타자에게 '형태적', '은유적',

I had much to worry about from being asked to criticize the architecture of UnSangDong. I have been watching UnSangDong for a long time, so I was cautious and it was difficult for me to find points to censure. After looking closely at their architecture, questions sometimes came to mind, but I could not come up with clear answers. So, I will put off answering these questions that I have not yet properly formed, and talk about the various misconceptions and prejudices regarding the architecture of UnSangDong. Some of these misunderstandings are follows. UnSangDong builds distinctive architecture that reacts instantly, rather than an architecture that creates a sound system by consistently incorporating architectural elements. There is a sense of difference between language and practice, in which the text containing UnSangDong's theories and philosophical thoughts seems to surround the buildings. These misunderstandings may to some extent be true. The architects of UnSangDong are not people who sincerely arranges themselves with a certain historical and discursive context. Many references and concepts quoted in the text are closer to the social attitudes they solve, rather than the worries of their work. An explanation of what they are making is on a different level from the work being made. These large and small refractions are sometimes a good target for making hasty judgements about "morphological,"

'표면적' 건축이라는 성급한 판단을 내기에 아주 적합한 타깃이다.
하지만 이런 오해로 인해 운생동에 대한 비평이 왜곡되거나 가벼움으로 보여지는 것은 적절하지 않다. 분명한 점은 한국 현대건축의 풍경에서 운생동이 자리하고 있는 지점이 매우 흥미롭다는 것이다. 아틀리에 사무소를 고집하면서 작은 규모의 프로젝트부터 상업건물, 주거, 공공, 도시 등 다양한 범위의 프로젝트를 수행해오고 있고, 전시, 출판, 교육 등의 범주도 활발히 참여한다. 아마도 오랜 기간 동안 운생동을 이렇게 살아남게 한 원동력은, 그들의 작업이 겉보기와는 다르게 숙련되고 능숙하게 실행으로 옮겨지는 그들만의 '실용(pragmatics)'에 근거하기 때문일 것이다. 현대건축의 시장에서 지명이 되고 주목이 되는 건축가로서 그의 영역을 만들어가는 것은 단지 뛰어난 재능만으로는 불충분하다. 건축가가 개인적으로 열망하는 조형적 탐색을 시장의 현실과 결부시키는 고유의 작업방식(modus operandi)을 지니고 그 결과와 논리가 계속 설득력을 지녀야 한다.[1] 운생동이 증명한 점은 (그 평가나 고정관념이 무엇이 되었건) 그들의 내용이 건축시장에서 충분한 가치를 발휘한다는 것이다.

운생동의 건축에서 겉으로 내보이는 특이함은 철저히 이 사무소를 이끄는 장윤규, 신창훈의 미학적, 문화적 기호와 관련이 있다. 논리보다 우선되는 어떤 감각으로 틀을 잡는다. 누구는 이것을 영감이라고 부르고 다른 이들은 작가정신이라고 칭한다. 약간 무모해 보일 수도 있지만, 새로운 건축적 질문을 찾기 위한 그들의 '놀이'이고, 놀이로 시작하지만 '치열'하게 접근한다. 그래서 운생동의 프로젝트가 추구하는 결정적인 뉴앙스는 장윤규, 신창훈의 스케치에서 그 구심점을 찾아볼 수 있다. 이 스케치들은 대부분 입면, 즉, 최종 건축물이 가지는 '인상'들을 가장 잘 포착해내서, 이 스케치를 보고 건물을 읽는 것과, 안 보고 읽는 것은 중요한 차이가 생긴다. 감각적인 스케치에 뼈와 살을 붙이듯 운생동 특유의 지식이 덧입혀지고 채워지는 방식. 이 특유의 실용은 주로 입면으로부터 점점 발전되는 과정으로 인해 단면까지는 살아 남을지언정, 많은 경우 평면의 희생 혹은 양보를 불러 온다. 그럼에도 불구하고, 평면이 최선을 다한다고 느끼는 이유는 내부공간이 최대한 성실하게, 효율적으로 구성되어 있기 때문이다. 그들의 건축이 '낯설게 하기'라는 전술임을 본인들이 밝히지만 평면은 낯설지 않고 매우 편안하다.[2] 감각적인 껍질과 단단한 알맹이, 물렁한 피부와 경직된 장기. 어울리지 않는 의아함이 여기저기서 드러난다. 만약 이것이 그 '낯설음'일까 하고 물어본다면 '그렇다'라고 답하기에는 설득력이 떨어진다. 몽유도원도 이상봉타워나 퓨처리즘 그리드 미동전자의 표피은 유려하고 출렁이지만 평면은 놀랍도록, (그리고 가끔은 허탈하게도,) 직선적이고 경제적이다. 외부와 내부에 대한 전혀 다른 태도는 운생동 특유의 이질감을 생성해낸다. 그리고 성실하게

"metaphorical," and "superficial" architecture.

However, it is not appropriate to distort or undervalue the reviews created by these misunderstandings. Clearly, UnSangDong's position in the modern Korean architectural field is very interesting. UnSangDong has been working on a vast group of projects ranging from small-scale commercial buildings to residential structures and public and urban spaces. The group is also actively involved in exhibitions, publishing, and education. Perhaps the driving force behind the long-term survival of UnSangDong is their ability to skillfully execute based on their own "pragmatics." It is not enough to make this area notable and remarkable in the contemporary architecture, just through excellent talent. The architects must have a modus operandi that connects personally aspirational and formative exploration with the reality of the field, and the resulting logic must be persuasive.[1] Fortunately, what UnSangDong has proved (whatever their assessments and stereotypes) is that their content has sufficient value in the field of architecture.

The uniqueness of UnSangDong's architecture is related to the aesthetic and cultural symbols of Yoon Gyoo Jang and Changhoon Shin, the architects who lead UnSangDong. They have set up a framework with a certain sense rather than logic. Some call it inspiration and others call it the spirit of the artist. It may seem a bit reckless, but they "play" to find new architectural questions, beginning playfully but approaching intensity. The decisive nuances that UnSangDong pursues can be found in the sketches of Yoon Gyoo Jang and Changhoon Shin. Most of these sketches capture the facade, the final impression of the building, so there is an important difference between looking at the building knowing the sketch, and doing the same but not knowing the sketch. UnSangDong's unique knowledge is added and filled like bones and flesh, all within that sensible sketch. Even though this peculiar practicality affects the gradual development of the process from facade to cross-section, it results in a sacrifice or concession of the plane. Nevertheless, I feel that the plane is doing its best because the inner spaces are constructed as effectively as possible. Although the architecture's tactics seem to include "making people feel strange," the plane is familiar and comfortable.[2] A sensitive shell, solid substance, soft skin, rigid organs, and unmatched wonders are revealed here and there. If you ask if this is "unfamiliar," it is difficult to answer yes. The surfaces of the Mongyudowondo Lie Sangbong Tower and futurism-based grid of the Midong Electronics and Telecommunication Headquarters are smooth and rolling, but the planes are surprisingly (and sometimes frustratingly) linear and economical. A totally different attitude towards the outside and inside creates a sense of heterogeneity that is unique to the universe. In addition, in the rigid attitude of the Korean architectural field, sincerely seeking connection with and harmony between the outside and inside, and the virtue of the architecture as the depth of the architectural system, the architecture of UnSangDong can be so polemic.

외내부와의 연계, 조화를 추구하고 건축의 미덕을 건축적 체계의 깊이라고 통용되는 한국건축의 경직된 써클에서 운생동의 건축은 그래서 문제적(polemic)일 수 있다.

현대도시의 조건과 현상에 대해 문제적임에도 불구하고 스스로 적응해 낸 운생동의 언어와 그 유희에 대해 그들의 작업을 어느 정도 존중한다. 동시에 그 이상이 있는가 하면 선뜻 답을 못 하는 것이 아쉽기도 하고 왜일까 궁금하기도 하다. 이 점이 특별히 아쉬운 이유는 그들이 버릇처럼 만들어 놓은 한계를 넘었던 경우가 종종 나타난다고 생각하기 때문이다. 운생동의 초기 작품 중 하나인 예화랑(2005), 그리고 운생동의 작품소개에 매번 첫 페이지로 등장하는 크링(2008), 그리고 실현은 안되었지만 논쟁적인 아시아문화전당 공모작 등에는 각각 그들의 실용이 새로운 담론의 영역으로 넘어갈 수 있는 교육적 잠재성(learning potential)이 있었다. 장윤규가 말하는 외피가 도시로 넘어오는 역할이 그 설득력을 갖출 수 있었고, 낯설지만 수긍하게 만드는 무언의 힘이 있는 듯 해 보였다. 표면에서 드러나는 어휘적 플레이가 도시의 조건과 맞아 떨어지는 것은 물론 그 조건에 대한 새로운 생산으로 반응했다. 수직으로 찢어진 예화랑의 외피적 제스처(gesture)는 빛의 조절이 중요한 갤러리의 요구사항은 물론 내부공간의 조직에도 영향을 미치면서, 파편화되어 있는 가로수길의 풍경에 '낯설음'을 선사한다. 틈틈이 벌어지면서 올라가는 수직 외벽의 띠들은 어느 각도에서 보느냐에 따라 백드롭(backdrop)이 되기도 하고 재질을 가진 커튼이 되며, 도시의 스크린이 되기도 한다. 길에서 보는 시야가 제한되어 있는 가로수길의 특성에 대해 과하지도 묵언하지도 않는 은근히 대담한 방식으로 말이다. 이와 비슷하게 크링은 운생동스러운 특유의 화려함 뒤에 그들이 모델하우스로 대변되는 시장의 특성을 현명하게 이용한 흔적이 숨어 있다. 겉모습은 경직된 대치동의 거리에 대해 최대한 압도적이고 세련된 모습으로 도시를 향해 갈구하는 반면, 내부공간에 배치된 다양한 공간적 변이는 친근하고 접근 가능하다. 아시아문화전당은 지표면이 들어올려짐으로써 새롭고 다르게 인식될 수 있는 도시적 시야와 도시-건축 사이의 유희가 있었다. 거대한 규모의 공공시설에서 느껴질 수 있는 압도감을 고르게 (하지만 역동적으로)분절시켜 공간에 대한 부담을 조절하고자 하는 의도도 엿보였다. 외부의 길과 내부의 길, 분절된 공간이 내부와 외부로 전환되는 방식이 전경-배경의 관계를 교란시켜 결국 해체시켜 버리는 것도 설득력이 있었다. 특유의 제스처는 여전히 있다. 하지만 이 프로젝트들은 고유의 상황 속에 숨겨진 잠재적 조건이 더해져 스토리의 힘이 배가된다면 그 논의는 무한확장이 가능해 질 수 있음을 시사했다고 생각한다. 그들의 이질적 실용에 거부감을 느끼는 다른 이들도 바로 수긍시킬 수 있는 정도의 힘 말이다.

Despite the fact that the conditions and phenomena of the modern city are problematic, I respect their work because of the language and amusement they adopted. The works offer more than I had expected, but at the same time I wonder why they cannot answer certain questions. This is especially true because I think that they often exceed the limitations they make, almost out of habit. Gallery Yeh (2005), one of the earliest works of UnSangDong and Kring (2008), which appears on the first page of the introduction to UnSangDong, as well as the controversial Asian Culture Complex, which has not yet been constructed, all have the potential to move forward into the domain of knowledge and discourse. The role of the surface that is transferred to the city described by Yoon Gyoo Jang was convincing and seemed to have an unfamiliar but acceptable power. The lexical play on the surface responds to the conditions of the city, as well as the new production based on those conditions. Vertically torn gestures of the surface of Gallery Yeh affects the organization of the interior space, as well as the requirements of the gallery where light control is important. Also, it gives strangeness to the landscape of the fragmented, tree-lined street. The vertical outer wall bands that rise as the gap spreads up become a backdrop, a curtain that has a certain material, and a city screen, all depending on the angle from which you looking. The tree-lined street, where the view from the road is limited, is gentle but has a bold manner. Likewise, Kring hides the traces of its wise use of the characteristics of the market, which it represents as a model house behind its unique splendor. The outward appearance of Kring is as intense and sophisticated as possible with regards to the city, but spatial variations arranged in the inner space are intimate and accessible. The Asian Culture Complex offers an amusing between urban vision and city architecture, which could be perceived as new and different since the surface of the land was lifted. The intention was to evenly (but dramatically) divide the overwhelming-ness that can be felt in large-scale public facilities in order to control the burden of space. It is persuasive in the way that the external and the internal roads and divided space are transformed into inside and outside in order to disrupt and deconstruct the relationship between the foreground and back. It is a distinctive gesture. However, I think that these projects suggest that if the potential conditions hidden in their own circumstances are added to the strength of the storyline, the discussion can be infinitely extended. Others who are reluctant regarding their heterogeneous pragmatism are capable of accepting it.

How has architecture played a role in the phenomena and conditions of the time? Modernism in the 20th century gave a fairly clear answer to this question. Architecture contributes to the public; it fulfills and helps to create a better society. An important clue came when architecture theorist Colin Rowe declared the end of modernism in his introduction to "Five Architects," which illustrated the break in this program of architecture and social.[3] Postmodernism explored the criticism of modernism

건축은 어떻게 시대에 따른 현상과 조건에 대해 그 역할을 담당해 왔을까. 20세기 모더니즘은 이 질문에 대해 꽤 명료한 답을 내놓았다. 건축은 공공에 기여하는 것이고 더 나은 사회를 만들어가는 책임을 수행하고 일조하는 것이다. 건축이론가 콜린 로우(Colin Rowe) 가 '다섯 명의 건축가들'(Five Architects)의 서론에서 모더니즘의 종말을 선언하게 된 가장 중요한 단서는 바로 이 건축과 사회적 프로그램의 단절이었다.[3] 포스트 모더니즘은 모더니즘에 대한 비판과 그로 인해 얻은 건축의 자율성을 탐색했다. 넘쳐나는 글과 비평, 토론, 담론으로써 건축이 논리적으로 명쾌한 해답을 찾아내는 오랜 모더니즘의 전통과 답습을 향한 고의적인 돌아가기였고, 도발이었다. 현재의 시간은? 어떤 이들은 오늘을 가리켜 수퍼모더니티(supermodernity)라고 부른다.[4] 장소를 근거로 하던 우리의 삶과 생활이 인터넷, 초고속 교통, 푸드마일, SNS와 같은 비장소(Non-place)가 주도하는 삶으로 변하는 시대. 국가나 기업 같은 실질적 주체 대신 추상적인 자본시장경제가 세계화를 가속화시키고 그 결과로 물질세계를 변화시킨다. 이 세대는 분명 보이고 만져지는 물질과 대상에 대해 논리적으로 접근하여 문제해결을 합리화하는, 혹은 의도적으로 그것을 반박하는 태도로 사고하여 바라보았던 과거의 세계관과는 다르다. 깊게 사고하여 탐색함으로써 이론을 구축하고 명분을 만들던 시대가 모더니티(포스트모더니티)라면, 수퍼모더니티에는 이 모든 원칙과 무관하게 다른 차원의 세계가 있다.

아마도 운생동이 작업을 통해 말하고자 하는 많은 부분은 수퍼모더니티라는 시대적 현상에 대해서의 고민일 것이다. 그들은 분명히 이 '다름'을 알고 있는 듯 하다. 운생동이 열중하는 도시적 행동은 건축내부공간의 문제보다 건축을 어떻게 드러내고 그 효과에 대해 가상적으로 음미할 수 있는 한 이미지를 만들어내는 문제와 연결되어 있지는 않을까. 더 독특하고, 더 새롭고, 더 화려한 이미지와 그를 구현해내는 기술이 가능한 세계에서, 건축을 보는 관점과 기준은 확연히 변화하고 있다. 건축을 하나의 체계로 바라보던 공간논리, 구축논리는 바로 이렇게 등장한 새로운 '감각'에 의해 덮여 쓰여진다. 건축을 만드는 요소와 체계는 변함없지만 그 요소와 체계가 만들어내는 영향력은 예전과는 다른 것이다.

그리고 이 다름에 대해 활발히 탐색하는 운생동의 이질적 실용은 아직까지는 유보적이다. 오히려 현대건축의 새로운 양상에 대해 우리 스스로가 어떻게 생각할 지 혼란스러운 이 상황에서, 건축에 대한 탐색을 놀이하고 배회하는 운생동의 건축은 차라리 솔직해 보인다. 하지만 이것도 부족하여 일말의 기대를 하고 있다. 만약, 드문드문, 그들이 자신의 영역 그 이상으로 가는 징후가 나타난다면, 그리고 그 징후들이 반복적이라면. 운생동의 건축이 가지는 실용 사이의 거리감은 어느 순간 그 스스로가 의미 있는 공간이 되지 않을까 한다.

and autonomy of architecture based on that criticism. It was a deliberate return to and provocation of the old tradition of modernism, an effort to find logically clear answers to architecture through overflowing writing, criticism, discussion, and discourse. What about our current time? Some call today super-modernity.[4] We are living in an era in which our lives used to be based on a place, but now are transformed into non-place lives by the Internet, high-speed transportation, food mileage, and social networking. An abstract capital market economy, instead of a substantive entity such as a country or corporation, has accelerated globalization and as a result changed the material world. This generation differs from the past worldview that rationalized the solution to the problem by logically approaching materials and objects clearly visible and touched, or intentionally refuted them. If modernity (or postmodernity) was the era of constructing a theory and making a cause by deeply thinking and exploring, super-modernity is a different world, separated from all of these principles.

Maybe UnSangDong's main point concerns the worries related to the era of super-modernity. They obviously seem to know the difference. Urban action, which UnSangDong emphasizes, can be linked to the problem of creating an image that can reveal architecture itself rather than the problem of interior spaces in architecture; it can be virtually enjoyed. In a world where more unique, newer, and more brilliant images and technology that enable them to be realized, the viewpoints and standards of architecture are changing drastically. Space and construction logic, which once looked at architecture as a system, is now covered by a new and emerging sense. The elements and systems that make architecture are unchanged, but the influences of those elements and systems are different.

Moreover, the heterogeneous practicality of UnSangDong, which is actively exploring the difference, continues to be reserved. In this confusing situation where we must consider a new aspect of modern architecture, the architecture of UnSangDong, which explores and roams, seems rather straightforward. However, this is not enough; I expect something more. If they sometimes show signs that they go beyond their territory and those signs are repetitive, the sense of distance between the practicalities of UnSangDong's architecture may at any moment create a meaningful space.

1 건축에서 실무는 특정한 조건, 상황, 요구와 작업을 해야 한다. 하지만 이 조건들은 건축가마다 풀어내고 대응하는 방식이 모두 다를 것이다. 수년을 작업하면서 얻은 경험도 중요하지만 막연한 것에 대해 최적치의 시나리오를 설정하는 것도 역시 건축가의 몫이다.
2 장윤규는 도시 풍경을 변화시키는 하나의 방식으로 기존의 컨텍스트에 반하는 어휘를 개입하는데서 출발한다고 밝히며 이를 낯설게 하기와 연결시킨다. 논리적 오류의 풍경, p.38, Space 609, 공간사
3 Colin Rowe, Introduction, Five Architects: Eisenman, Graves, Gwathmey, Hejduk, Meier, MOMA, 1972
4 프랑스 인류학자 마크 오제(Marc Auge)는 수퍼모더니티를 과잉 정보, 과잉 공간, 과잉 시간 등의 '과잉'의 시기라고 정의한다. 넘쳐나는 정보, 보이지 않는 공간, 보이는 대상이 없이도 무엇인가가 만들어지고, 제조기술은 정보기술로 대체되는 시대를 가리킨다. Marc Auge, Non-Places: An Introduction to Supermodernity, Verso, 2009

1 Practice in architecture should have specific conditions, circumstances, demands, and tasks. However, the solutions and countermeasures will be different depending on the architects. The experience gained from working for many years is important, but it is also up to the architect to replace this vague scenario with an optimal one.
2 Yoon Gyoo Jang says that it begins with the intervention of the vocabulary, contrary to the existing context, as a way of changing the urban landscape. He connects it with the unfamiliarity. Landscape of Logical Error, p. 38, Space 609, SPACE
3 Colin Rowe, Introduction, Five Architects: Eisenman, Graves, Gwathmey, Hejduk, Meier, MOMA, 1972
4 French anthropologist Marc Auge defines "super-modernity" as a period of surplus, such as in excessive information, space, and time. If refers to an era in which something is created without overflowing information, invisible space, and visible objects, and manufacturing technology is replaced by information technology. Marc Auge, Non-Places: An Introduction to Supermodernity, Verso, 2009.

Jeongok Prehistory Museum
전곡리 선사박물관

Location: 528-1 Jeongok-ri, Yeonchun-gun, Gyeonggi-do, Republic of Korea **Architects:** Jang Yoon Gyoo, Shin Chang Hoon **Designer:** Kim Woo Young, Kim Yoon Soo, Kim Sung Min, Choi Hye Jin, Kwon Woo Seok, Jung Bok Joo **Photographer:** Jaekyeong Kim(Model)

지형은 자연의 현상을 대변하는 결과물인 동시에 시간의 흐름에 반응하는 생물체이다. 전곡리에서 지형은 건축의 대상지이면서 선사박물관을 구축하는 프로그램적 요소이며, 형태적 모티브가 된다. 표피적으로 지형은 자연시간의 축척된 현상으로 설명될 수 있다. 그러나 여기선 지형을 인간 문명활동과 인류의 역사적 현상을 담는 본질적 요소로서 설정하려한다. 지형을 통하여 고대의 시간과 현재의 시간, 미래의 시간을 연결하는 인류문화의 유동적이며, 진화하는 인류에 필요한 여러 프로그램을 구체적으로 정의한다. 지형은 야외전시장, 선사체험장, 조경공간, 이벤트전시장, 현무암 단층전시대, 한탄강전망대 등으로 진화된다. 개별적인 지형의 진화는 '시간의 복도'라는 네트워크 프로그램을 통하여 선사박물관 구축의 기본구조 역할을 할 것이다. 이는 단순한 선사시대의 역사적 박물관을 넘어서 현 인류의 다양성과 무수히 다변화되는 요구 속에서 반응하는 진화하는 생명체로서 열린 선사박물관이 될 것이다.

We propose creating a museum program and space by trans-morphing the earth. The outcome would represent the evolution of nature while the body acts as the change of time. The natural topography of Jeongok-ri is a major element of the site as well as its formal motif and programmatic constituency. Through topography, programs can be defined and prescribed for our past, present, and future in conjunction with mankind's fluid and ever evolving cultural landscape. The topography will evolved into the outdoor exhibit by establishing an area for a prehistoric experience, open space, event hall, a basalt precipice exhibition, Observatory over the Hantan river, etc. Through the Time Corridor, the evolution of each space will form a network and become a framework for the entire museum. Reaching beyond the idea of the museum, which focuses on prehistory, we hope to enervate the museum, creating a living, dynamic body that reacts and breathes with the thriving diversity of mankind and changing demands.

Hannae Forest of Wisdom
한내 지혜의 숲

Location: Seoul, Republic of Korea **Architects:** Yoongyoo Jang, Changhoon Shim **Design team:** Soonhoon Choi **Client:** Nowon-gu District Office **Completion:** 2017.03 **Gross Floor Area:** 359.37m² **Photographer:** Jaekyeong Kim(Model), Junhwan Yoon, Sergio Pirrone

한내 지혜의 숲은 거대한 공공 개발 프로젝트에 의해 만들어진 문화 공간이 아니다. 그것은 사람들이 삶을 즐기고 다시 돌아오기를 바라는 창조적인 공간에 대한 아이디어로 시작되었다. 그래서 우리는 버려진 도시 공간을 창의적 아이디어를 통해 독특하고 매력적인 공간으로 만들었다. 운생동의 건축가들은 한내 지역 문화 재건을 위한 작은 공동체 공간을 제안한다. 330qm 미만의 바닥면적을 가진 작은 도서관이지만 내부에서 방문객은 다양한 공간적 측면을 경험한다. 이 도서관은 크기 한계를 극복하고 다양한 조건을 통해 다기능 공간으로 발전한다. 박공지붕은 벽(서가 선반)의 연속이며 다양한 높이의 문지방을 형성한다. 이 지점을 통해 자연광이 안으로 떨어진다. 미로는 도서관을 거니는 아이들의 상상력과 창의력을 자극한다.

Hannae Forest of Wisdom is not a cultural space generated by a big public development project. It was initiated by the idea of a creative space, where people enjoy to inhabit and willing to come back every day. Therefore we improved an abandoned urban space into a unique and attractive space with creative ideas. Unsangdong Architects offers a small community space for Hannae area, which can regenerate the local culture.
This is a small library with the floor area under 330qm, but the visitors experience various spatial aspects inside. The library overcomes its limitation in terms of size and develops itself into a multi-functional space through different conditions. The gable roof is the continuation of the wall(book shelf) and forms threshold through varying heights. Natural light falls inside through this point. The labyrinth stimulates the imagination and creativity of children meandering through the library.

Lie Sang Bong Tower
이상봉 타워

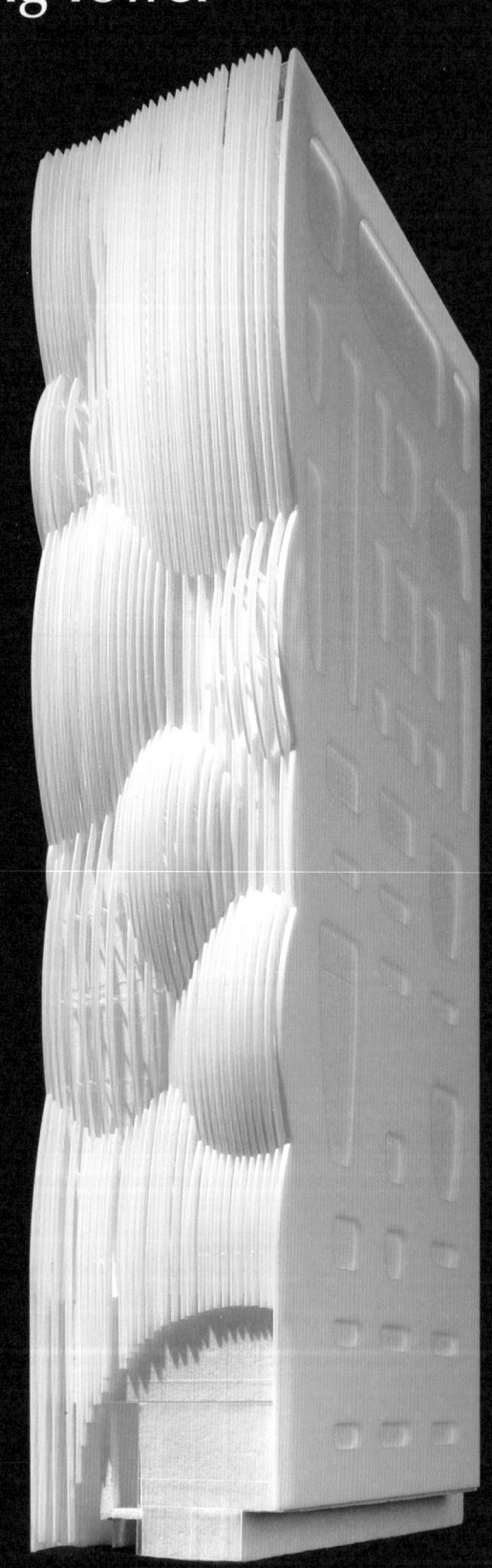

Location: 97-6, Cheongdam-dong, Gangnam-gu, Seoul, Republic of Korea **Use:** Commercial, Office **Gross Floor Area:** 5,007.34m² **Building scope:** B5 ~ 14F **Height:** 69.97m **Architects:** Yoongyoo Jang, Changhoon Shin **Design team:** Bongkyun Kim, Minseung Moon, Huijin Lee, Minkyun Kim, Taekmin Kim, Inhu Lim, Yeseul Jeon, Minji Sohn, Yuyeong Hwang **Structure:** Reinforced concrete structure **Client:** Lee Sangbong **Photographer:** Yongkyeong Kim(Model), Jaeyoun Kim

일반적인 건축의 스킨을 배재하고, 동양 산수 특히 몽유도원도적인 산수를 건축에 맵핑하려 한다. 이는 마치 몽유도원도가 꿈속의 풍경을 재현했듯이 자연으로 발췌된 다양한 추상적인 도형의 결합을 상상속의 이미지로 결합하여 새로운 산수화와 같은 스킨으로 변화시키려 한다. 모더니즘의 표상이라 할 수 있는 직교체계의 좌표를 해체하고 변형시킴으로서 직교자표 틈새에 숨어있던 공간과 구조를 새로운 시각으로 발견해낸다. 곡선형 프레임화된 단면적 스킨은 내부와 외부의 모호한 경계와 틈새를 만들며 영속적인 형상과 그것의 감각적인 사본 사이 존재하는 제3의 유(Genos)위에 각인되어있는 장소를 창조한다. 비어있는 듯이 보이지만 비어있는 것이 아니고 감각적인 세계라는 의미에서 일시적인 것도, 형상이라는 의미에서 영속적인 것도 아닌 모든 것들이 자리잡을 수도 있고 발생될 수도 있으며 또한 각인될 수도 있는 틈새(spacing)의 건축을 발견하는 것이다. 우리는 이상봉 디자이너와 몇 년간 협업할 기회를 찾아보자고 이야기해 왔었다. 이상봉 디자이너와 운생동은 라이프스타일에는 분명한 경계가 없으며, 건축이 곧 패션이고, 패션이 곧 건축이라는 데 공감해 왔다. 이상봉 타워가 들어서는 청담동은 유명한 해외 패션브랜드들의 플래그십스토어로 가득찬 곳인데, 이 곳에 자리잡는 한국 디자이너의 건물은 어때야 할까에 대해 고민을 많이 했다. 건물의 디자인은 한국적인 미학을 현대적으로 해석하는 이상봉 디자이너의 비전이 반영되었다.

Lie Sang Bong is Korea's leading fashion designer who creates contemporary style reflecting traditional Korean and Oriental aesthetics. He is work is famous for getting a motif and inspiration from Korean alphabet Hangeul. He does shows in New York Fashion Week and Paris Fashion Week. Unsangdong and Lie Sang Bong have been discussing collaboration for several years as they share the vision that "fashion is architecture and architecture is fashion". They both believe there is not a clear division among different sectors when it comes to forming a lifestyle. Lie Sang Bong Tower is an iconic additional to Cheong-dam Dong, which represents a posh Gangnam area in Seoul. The building is right next to luxury clothing brand Shingsegae International's HQ designed by Olson Kundig. Unsangdong was conscious that this is a rare flagship store by a Korean designer in an area which is otherwise packed with foreign high-end designer flagship stores from House of Dior Boutique by Christian de Portzamparc and Peter Marino, Givenchy HQ designed by Piuarch. Lie Sang Bong Tower houses Lie's showroom and studio from the ground floor to the 2nd floor. Above the studio are the office rental (4th-8th) and premium residence space (9th-13th floor). The top two floors (14th-15th) are devoted to cultural space managed by Lie Sang Bong as a gathering place and will be widely open for fashion and creative people. Unsangdong is discussing various cultural programs with Lie.

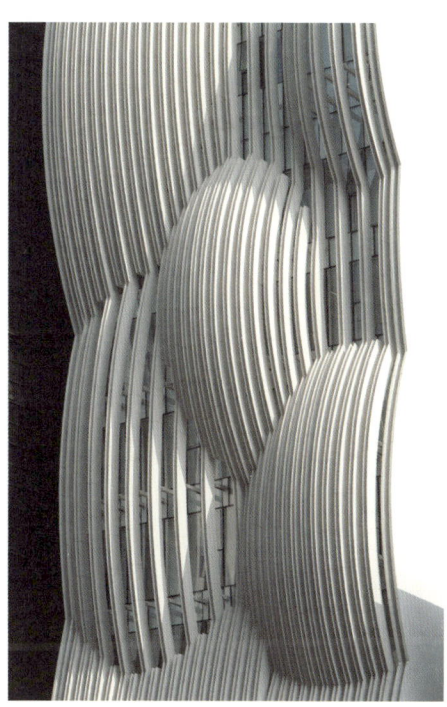

Profile

운생동 UNSANGDONG ARCHITECTS

장윤규 Yoongyoo Jang

건축가 장윤규, 서울대 건축과 및 동 대학원을 졸업했다. 현재 '운생동건축가그룹' 대표, 국민대 건축대학 교수, '갤러리정미소' 대표로 활동하고 있다. 주요 건축 작품으로 금호복합문화공간 크링, 예화랑, 서울대 건축대학, 홍익대 대학로캠퍼스, 파주출판단지의 생능출판사, 광주 디자인센터, 이집트 대사관 등이 있다. 세계적인 건축상인 AR Award 와 뱅가드상을 비롯해 2008년 한국공간디자인대상 대상, 대한민국 우수디자인(GD) 국무총리상, 서울시건축상 등을 받았다.

신창훈 Changhoon Shin

건축가 신창훈, 건축가 그룹 운생동의 공동대표, 영남대학교 건축공학과를 졸업하고 서울시립대학교 건축대학원을 졸업하였다. 아르텍건축, 범건축, 힘마건축에서 실무 경험을 쌓았다. 장윤규와 함께 실험건축, 개념적 건축을 실현하기 위해서 건축가 그룹 운생동을 결성하여 예화랑, 서울시립대 종합강의동, KRING, 성수문화복지회관, 청심물문화관 등 다각적 작업을 진행하고 있다. 현재 서울시립대 건축학과 겸임교수, 서울시 공공건축가로 활동하고 있다.

다니엘 바예 Daniel Valle
중앙대학교 건축학과 Department of Architecture, Chung Ang University

다니엘 바예는 마드리드의 E.T.S.A.M를 수석으로 졸업하고 네덜란드 Berlage Institute에서
건축학 석사 학위를 취득했다. 이후 런던, 마드리드, 서울, 아부다비에서 근무했으며,
2008년 스페인 마드리드에 Daniel Valle Architects를 설립했다. 지난 15년 간
전 세계 많은 대학에서 건축 디자인을 가르쳤으며, 현재는 중앙대학교에서 강연을 하고 있다.

김현섭 Hyonsob Kim
고려대학교 건축학과 Department of Architecture, Korea University

영국 셰필드대학교에서 서양 근대건축을 공부했고, 2008년 모교인 고려대에 임용된 이래
건축역사·이론·비평의 교육과 연구에 임하고 있으며, 근래에는 한국 현대건축에 대한
비판적 역사 서술에 관심을 모으고 있다. 최근『건축수업: 서양 근대건축사』(2016),
『건축을 사유하다: 건축이론 입문』(2017),「DDP Controversy and the Dilemma of H-Sang
Seung's "Landscript"」(2018),「르 코르뷔지에와 한국의 현대건축」(2018) 등을 출판했다.

박진호 Jinho Park
인하대학교 건축학과 Department of Architecture, Inha University

박진호 교수는 UCLA 건축대학원에서 건축이론 분야 석사 및 박사 학위 취득 후,
미국 하와이주립대학교 건축대학에서 조교수 및 부교수(종신교수)를 역임하였다.
세계 유수의 건축저널에 다수의 논문을 발표하고 있고,
최근 저서로는 Images Publishing에서 출간된 "Graft in Architecture: Recreating Spaces" (2013),
"Designing the Ecocity-in-the-Sky" (2014) 등이 있다.

백승만 Seungman Baek
영남대학교 건축학부 School of Architecture, Yeungnam University

한양대학교 건축학과를 졸업하고, 파리 벨빌 국립건축대학(ENSA Paris-Belleville)에서
디플롬, 파리 라 빌레뜨 국립건축대학(ENSA Paris-La Villette)에서 포스트디플롬,
파리 사회과학고등연구원(EHESS)에서 박사학위를 취득하였다. 서울(시정개발)연구원, 공간,
정림 등에서 설계실무를 경험하였으며, 프랑스정부공인건축가(DPLG)로서 2010년부터
영남대학교 건축학부에서 전임교원으로 재직하고 있다. 영주시 공공건축가,
재한 프랑스건축사회장 등을 맡았으며, 현재 (사)한국건축설계학회 부회장,
(사)한국건축가협회 미래건축위원회 위원장을 맡고 있다.

송하엽 Hayub Song
중앙대학교 Chung-Ang University

송하엽은 서울대 건축학과를 졸업하고 건축설계 실무 경험을 쌓았다.
이후 미시간 대학교에서 건축학 석사 학위를, 펜실베이니아 대학교에서 박사 학위를 받았다.
" 22세기 건축", 『랜드마크; 도시들 경쟁하다』, 『파빌리온, 도시에 감정을 채우다』(공저)
『전환기의 한국 건축과 4.3그룹』(공저)을 쓰고, 『표면으로 읽는 건축』을 우리말로 옮겼다.
주요 디자인 작품으로 [U_GROWING PARK] [바람 같은 돌]이 있으며, 2014~2016년 서울건축문화제에서
'담박소쇄노들: 여름건축학교' '한강감정: 한강건축상상전' '한강힌트: 한국건축상상전'을 기획했다.
현재 중앙대 교수로, '서울공예박물관 설계' '수상레포츠통합센터 설계' 등을 작업 중이다.

안기현 Keehyun Ahn
한양대학교 건축학부 School of Architecture, Hanyang university

안기현은 한양대학교 건축학부 교수로 재직 중이다.
한양대학교와 미국 U.C. 버클리대학교 졸업하였고, 한국, 미국, 유럽에서 실무를 익힌뒤,
2010년에 AnLstudio를 설립하여 실험적이고 다양한 스케일의 작업을 이어오고 있다.

염상훈 Sanghoon Youm
연세대학교 건축공학과 Department of Architecture & Architectural Engineering, Yonsei University

---

서울대학교 건축학과와 뉴욕 컬럼비아 건축대학원을 졸업하고
뉴욕과 유럽 및 국내에서 활동을 하고 있으며, 현재 연세대학교 건축공학과 교수로 재직하고 있다.
CAT건축도시디자인연구실을 운영하며 도시적 관점을 반영한 건축디자인과 더불어
디지털 기술에 대한 다양한 실험과 전략을 연구하고 있다. 건축의 정성적 가치의 정량화에 대한 연구와 함께
재개발과 재사용 건축에 대한 대안을 모색하고 있으며 기하학의 공간적 가능성과 설계방법론 및
건축교육방법론에 대한 연구를 수행하고 있다.

윤정원 Jungwon Yoon
서울시립대학교 건축학부 Department of Architecture, University of Seoul

윤정원은 서울시립대학교 건축학부 부교수로 재직하고 있으며,
프로젝트 기반의 건축 설계 작업을 진행하면서, 건축 재료 및 구축 방식을 중심으로 기술 및
영역 확장을 탐구하고 있다. 서울대학교와 미국 프린스턴대학교에서 건축을 공부하였고,
미국 뉴욕의 RMJM과 Rafael Vinoly Architects PC, 그리고 네덜란드의 OMA에서
다양한 건축 설계 프로젝트를 수행하였다. 미국 건축사, 한국 건축사 및 LEED BD+C를 취득하였고,
2016년부터 서울시 공공건축가로 활동 중이다.

이경선 Kyungsun Lee
홍익대학교 건축학과 Department of Architecture, Hongik University

———

이경선은 홍익대학교 건축학과와 UCLA 건축설계 석사(M. Arch I)를 졸업하고 하버드대학에서 친환경 건축계획으로 건축학 박사학위를 취득하였다. 석사 졸업 후 Moore Ruble Yudell, HLW International, Gwathmey & Siegel Associates등에서 다양한 건축실무 경험을 하였다. 현재, 홍익대학교 건축대학 교수로 재직하면서, 건축연구소 SUNe. lab을 이끌고 있다. 인간과 환경, 건축이론과 실무의 접목을 통한 과학적인 방법의 건축통합설계에 대해 관심을 가지고 "지속가능한 건축과 도시", "감성디자인", "아동과 청소년의 문화 및 교육공간" 등에 관한 지속적인 연구와 작품활동을 하고 있다.

이영범 Youngbum Reigh
경기대학교 건축학과 Department of Architecture, Kyonggi University

———

서울대학교 건축학과 및 동 대학원을 졸업하고 영국 AA Graduate School에서 Postgrad. Dploma와 Ph. D를 취득했다. 창조건축사사무소를 거쳐 현재 경기대학교 건축학과 교수로 재직중이다. 참여디자인, 마을재생, 공유공간, 공동체성, 시민자사산화 이슈에 관심을 갖고 이론과 현장을 넘나들며 활동하고 있다. 도시연대란 시민단체에서 공간운동을 지속적으로 하며 비영리법인인 도시와 삶의 이사장을 맡고 있다. 도시의 죽음을 기억하라, 뉴욕 런던 서울의 도시재생 이야기, 커뮤니티 디자인을 하다 등 다수의 저서가 있다.

이은석 Eunseok Lee
경희대학교 건축학과 Department of Architecture, Kyunghee University

———

이은석 교수는 홍익대 건축학과와 파리 국립 제 1 대학 소르본느를 졸업하고 예술사학 박사학위를 취득했다 . 국립 파리 벨빌 건축대학을 졸업한 프랑스 공인 건축사이며, 현재 경희대학교 건축학과 교수이다. 1995년 LA 한미문화예술센터 국제현상과 2000년 "천년의 문" 현상설계에서 1등으로 당선한 바 있다. 1996년 아뜰리에 KOMA를 개소한 이래, 주요 작품으로는 "새문안교회" 및 "하늘보석교회", "범어교회", "늘샘교회", "경산교회" 등 다수의 교회와 "부전 글로컬 비전센터", "총신대 신관 및 도서관", "꿈의 학교", "금성 유치원", "뱅루즈"등의 교육·문화시설, "리안주택", "청담 엔트라호텔" 등의 주거·상업시설까지 다양한 프로젝트를 진행하였다. 그의 저서로는 "아름다운 교회건축"과 "미완의 근대성"을 비롯한 여러 건축과 스케치의 작품집이 있다.

이 황 Hwang Yi
아주대학교 건축학과 Department of Architecture, Ajou University

———

이황은 서울대학교 건축학과를 졸업하고, 펜실베니아 대학 (University of Pennsylvania) 디자인 스쿨에서 "Information in Environmental Architecture" 라는 제목의 논문으로 박사학위를 받았다. 최적화 알고리즘, 환경 시뮬레이션, 로보틱스 등의 공학적 접근을 통한 정보기반의 디자인 방법론 개발 및 기술과 디자인 융합에 관심을 갖고 연구중이다. 김중업 및 김태수 장학제에 선정된 바있으며, 2016년 부터 미국 마이애미의 Florida International University (FIU) 건축대학 교수를 역임하였다. 현재 아주대학교에서 Design Engineering & Robotics for Sustainable Architecture (DEERS-Arch) 연구실을 이끌고 있다. 저/역서로 친환경 건축설계 시뮬레이션 (2009), 불완전한 건축 (2012) 등이 있다.

전유창 Youchang Jeon

아주대학교 건축학과/ aDlab+ 공동대표 Department of Architecture, Ajou University / Partner, aDlab+

전유창은 인하대 건축공학과를 수석졸업(용마루상)하고 미국 Columbia University에서
건축학 석사학위를 받았다. 1999-2007년까지 뉴욕 Mitchell/Giurgola Architects 에서 디자이너 및 이사로
재직 하였으며 35회 일본 Central Glass 공모전 대상을 및 다수의 국제 공모전에서 수상했다.
2010년부터 aDlab+ 디자인 연구소의 공동 대표로 한강 파빌리온(2011), 구로어린이집(2014),
사마르칸트 직업 훈련원(2015), 캄보디아 아클레다 대학 마스터플랜 및 설계(2016) 한전 에너지 센터(2018)
등의 프로젝트를 수행했다. 2007년부터 아주대학교 건축과 교수로 재직 중이며 미국건축사 및
미국 친환경 건축 인증사이다. 현재 서울시 공공건축가로도 활동중이다.

정만영 Mannyoung Chung

서울과기대 건축학부 School of Architecture, SeoulTech

서울시립대에서 건축을 전공한 후 같은 대학에서 '건축형태의 자의적 생성에 관한 연구'로
박사학위를 받았다. 일건건축에서 실무를 배웠다. 건축설계와 이론이 매개되는 지점에 서기를 원하며,
현대건축을 대상으로 실험, 외부성, 응축적 비평에 관심을 집중하고 있다.
2003년부터 10년간 철학아카데미에서 여름/겨울 건축강좌를 진행했으며,
한국건축역사학회 부회장, 한국건축가협회 편찬위원장을 역임했다.

조항만 Hangman Zo

서울대학교 건축학과 / 탈건축 DAAE, Seoul National University / TAAL ArchitectsUniversity

대한민국 건축사인 조항만은 서울대학교와 뉴욕의 컬럼비아대학교에서 수학하였고
서울의 ㈜ 경영위치 건축사사무소, ㈜ 아이아크 건축사사무소와 뉴욕의 GreenbergFarrow Architecture,
H Architecture 를 거쳐 현재 서울대학교 건축학과의 부교수로 재직 중이다.
또한 파트너 서지영과 함께 TAAL Architects를 설립하여 작품활동과 연구활동을 병행하고 있다.

최혜정 Helen Hejung Choi

국민대학교 건축학과 School of Architecture, Kookmin University

국민대학교 건축학과 조교수로 미국 렌슬리어 공대 건축학사(B. Arch), 콜럼비아대학원
건축디자인석사(MS. AAD)를 취득했다. 미국건축사로 활동 중 서울로 이주하여 건축가, 교수,
큐레이터로 활동하고 있다. 2011 광주 디자인비엔날레, 2017 서울 도시건축비엔날레 큐레이터를
역임했으며, 2014 광주아시아문화전당 문화정보원 책임연구원으로 활동했다.

Project Credit

크링(금호복합문화공간) KRING (KumHo Culture Complex)

Location: 9968-3 Daechi-Dong, Gangnam-Gu, Seoul, Republic of Korea
Use: Temporary Building
Building Scope: 3F
Site Area: 4,110.9m²
Building Area: 3,153.58m²
Total Floor Area: 7,144.53m²
Building Coverage Ratio: 76.71%
Floor Area Ratio: 173.06 %
Structure: Steel Structure+Stainless Steel, Sandwich Panel
Principals in Charge: Jang Yoon Gyoo, Shin Chang Hoon, Kim kyeung Tae
Design Team: Kim Sung Min, Moon Sang Ho, Kim Se Jin, Kang Seung Hyeun, Kim Bong Kyun, Goh Young Dong, Yi Na Ra
Client: KumHo E&C
Photographer: Jaekyeong Kim(Model), Sergio Pirrone

파주 생능출판사 LIFE & POWER PRESS

Location: 507-12 Munbal-li, Kyoha-eup, Paju City, Republic of Korea
Use: Office
Building Scope: B1F ~ 7F
Site Area: 727.20m²
Building Area: 352.50m²
Gross Floor Area: 995.72m²
Building Coverage Ratio: 48.47%
Gross Floor Ratio: 135.24%
Structure: R.C
Principals in Charge: Jang Yoon Gyoo, Shin Chang Hoon
Client: Kim Seung Ki
Photographer: Namgoong Sun

파리 올림픽 메모리얼 Paris Olympic Memorial

Location: Paris, France
Use: Landmark + Gallery
Structure: Reinforced plastic cell
Exterior finishing: Plastic wall, Full-color LEDs
Structure: Reinforced plastic cell
Photographer: Unsangdong(Model)

청심 물문화관 Cheongshim Purification Center

Location: 555-9 SongSan-li Seolak-myeon Gapyung Kyeunggi-do, Republic of Korea
Use: Purification Center
Building area: 505.00 m²
Gross Floor Area: 637.40 m²
Building Scope: B1 ~ 2F
Structure: R.C.
Photographer: Jaekyeong Kim(Model), Fernando Guerra, Sergio Pirrone

2012 여수 엑스포 현대자동차 그룹관 2012 Yeosu EXPO, Hyundai Motor Group Pavillion

Location: Site in Expo 2012 Yeosu, Jeonnam, Republic of Korea
Site Area: 1,960 m²
Building Area: 1,397.50 m²
Total Floor Area: 2,334.81 m²
Designer: Kim Sung Min, Kim Min Tae, Hyun Sang Heon, Goh Young Dong, Kang Seung Hyun, Jang Chol Min, Kim Mi Jung, Son Min Sun, Kim Hye Soo
Client: Innocean Worldwide
Structure: Steel Frame Construction
Photographer: Jaekyeong Kim(Model), Sergio Pirrone

크로노토프 월 하우스 House ONE : Chronotope Wall House

Location: Pangyo-dong 571, Bundang-gu, Seongnam-si, Gyeonggi-do, Republic of Korea
Completion: 03. 2016
Use: House
Site Area: 255.2m²
Building Area: 125.65m²
Building Coverage ratio: 49.23%
Gross Area: 228.78m²
Building Scale: B1 ~ 2F
Structure: Reinforced Concrete
Height: 11m
Design team: Kim Mi Jung, Kim Min Kyun
Photographer: Jaekyeong Kim(Model), Sergio Pirrone

예화랑 Gallery Yeh

Location: 532-9 Sinsa-dong, Gangnam-gu, Seoul, Republic of Korea
Use: Mixed-use facility
Site area: 567.5m²
Building coverage ratio: 58.62%
Gross floor area: 1,995.14m²
Scale: B2 ~ 7F
Construction: Gujin Industrial Development Co.,Ltd
Structure: R.C Client: Lee Sook Young
Exterior finishing T50 base panel, Exposed concrete, T24 transparent pair grass
Interior finishing: Epoxy coating, Base panel, Exposed concrete
Photographer: Jaekyeong Kim(Model) Yongkwan Kim

갤러리 더 힐 Gallery The Hill

Location: 810, Hannam-dong, Yongsan-gu, Seoul, Republic of Korea
Building Area: 485.076 m²
Total Floor Area: 2,730.31 m²
Architects: Jang Yoon Gyoo, Shin Chang Hoon, Kim Kyeong Tae
Designer: Sim Jehyun, Choi Young Eun, Ahn Bo Young, Yang Ki Ran
Client: Kim Sang Woon_Hansjaram Corporation
Structure: RC Structure
Photographer: Jaekyeong Kim(Model), Sergio Pirrone

오션어스 해운대 사옥 OCEANUS GROUP Haeundae Office

Location: 1502-10 Jung-dong, Haeundae, Busan, Republic of Korea
Site Area: 1,294.60 m²
Building Area: 771.29 m²
Total Floor Area: 2,996.36 m²
Architects: Jang Yoon Gyoo, Shin Chang Hoon
Designer: Kim Sung Min, Kang Seung Hyun, Kim Jung Seop, Ko Young Dong, Jang Chul Min, Ko Eun Jin
Client: Ocean Us
Structure: Steel Reinforced Concrete Structure
Photographer: Jaekyeong Kim(Model), Sergio Pirrone

코오롱 E + Green House Kolong E + Green House

Location: Jeondae-ri, Pogok-eub, Cheoin-gu, Gyeonggi-do, Republic of Korea
Site Area: 5,525 m²
Building Area: 957.40 m²
Total Floor Area: 1,837.04 m²
Architects: Jang Yoon Gyoo, Shin Chang Hoon
Designer: Kim Yoon Soo, Choi Young Eun, Ahn Hye Joon, Kim Ho Jin, Seo Hye Lim, Kim Mi Jung, Kim Ji Hye
Client: Kolon Global Corporation
Structure: RC Structure
Photographer: Jaekyeong Kim(Model), Sergio Pirrone

안성 남사당공연장 창고 Theater Contour (Namsadang)

Location: Bogae-myeon, Anseong-si, Gyeonggi-do, Republic of Korea
Site area: 8,264.5 m²
Building area: 2,411m²
Gross floor area: 3,147.19m²
Building to land ratio: 31.56%
Floor area ratio: 36.51%
Building scope: B1, 2F
Height: 20m
Photographer: Jaekyeong Kim(Model)

광주 아시아 문화전당 Asian Culture Complex in Gwangju

Location: Dong-gu, Gwangsan-dong, Gwangju, Jeonnam, Republic of Korea
Site Area: 93,375 m²
Building Area: 65,213 m²
Total Floor Area: 146,672 m²
Architects: Jang Yoon Gyoo, Shin Chang Hoon, Kim Woo Il, Kim Woo Young
Designer: Yeon Kyong Hee, Kim Yoon Soo, Kim Seong Min, Choi Hye Jin, Kwon Woo Seok, Jung Bok Ju
Photographer: Jaekyeong Kim(Model)

KTNG 복합문화센터 KTNG Culture Complex

Location: Paldal-gu, Suwon-si, Gyeonggi-do, Republic of Korea
Site Area: 7,504.30 m²
Building Area: 3,945.66 m²
Total Floor Area: 28,934.62 m²
Structure: RC Structure
Architects: Jang Yoon Gyoo, Shin Chang Hoon
Photographer: Youngkwan Kim(Model)

성동문화복지센터 Seongdong Cultural & Welfare Center

Location: 656-323, Seongsu-dong, Seongdong-gu, Seoul, Republic of Korea
Site Area: 2,204 m²
Building Area: 1,014.69 m²
Total Floor Area: 9,558.75 m²
Architects: Jang Yoon Gyoo, Shin Chang Hoon
Designer: Kim Sung Min, Kim Min Tae, Seo Hye Lim, Ryu Sam Yeol, Ahn Hye Joon,
Kim Won Il, Ahn Boo Young, Sim Jehyun, Kim Mi Jung, Jo Eun Jung
Client: Municipality of Seongdong-gu
Structure: Steel framed reinforcement concrete
Photographer: Jaekyeong Kim(Model), Sergio Pirrone

하이 서울 페스티벌 무대 조각 Hi Seoul Festival Stage Sculpture Palace pf May_Thousand Palace

Location : Seoul Plaza, Seoul, Republic of Korea
Allied Building around the site : Seoul City Hall, Seoul Plaza Hotel, JEI
Architects : UnSangDong Architects Cooperation
Principals : Jang Yoon Gyoo, Shin Chang Hoon
Co-Artist: Ahn Eun Mi
Structural Engineer : MakMax Korea
Materials : 3D MAK MESH(2 layered fabric made of translucent vinyl fiber)
Client : Seoul Metropolitan Government, Seoul Foundation Arts and Culture
Photographer: Sergio Pirrone

파주출판도시 어린이집 White Cube Matrix: Paju Kindergarten

Location: 525-4 Muoonbal-dong, Paju, Gyeonggi-do, Republic of Korea
Use: Nursery School
Site area: 1,120.3m²
Building Area: 495.62m²
Gross Floor Area: 1,009.34m²
Building to Land Ratio: 44.24%
Gross Floor Ratio 90.10%
Building Scope: 3F
Structure: R.C.
Completion: 2014. 7
Architects: Jang Yoon Gyoo + Seo Hyeon
Design Team: Choi Young Chul, Kim Mi Jung, Koh Eun Jin, Yoon Ji Soo
Photographer: Jaekyeong Kim(Model), Sergio Pirrone

잠실 종합 운동장 재개발 국제교류복합지구 A Thousand City Plateaus

Location: Bogae-myeon, Anseong-si, Gyeonggi-do, Republic of Korea
Site area: 8,264.5 m²
Building area: 2,411m²
Gross floor area: 3,147.19m²
Building to land ratio: 31.56%
Floor area ratio: 36.51%
Building scope: B1 ~ 2F
Height: 20m
Photographer: Jaekyeong Kim(Model)

여수 엑스포 2012 Yeosu Expo 2012

Architects : Jang Yoon Gyoo, Shin Chang Hoon, Kim woo Young, Kim Bong Kyun, Kang Seong Hyun
Photographer: Jaekyeong Kim(Model)

상하이 엑스포 2010 Shanghai Expo 2010

Location: Paldal-gu, Suwon-si, Gyeonggi-do, Republic of Korea
Site Area: 7,504.30 m²
Building Area: 3,945.66 m²
Total Floor Area: 28,934.62 m²
Structure: RC Structure
Architects: Jang Yoon Gyoo, Shin Chang Hoon
Photographer: Jaekyeong Kim(Model)

몽유도원도(상상의 도시) Mogyudowondo(Imagination City)

Architects : Jang Yoon Gyoo, Shin Chang Hoon + Mak Max Korea
Photographer: Jaekyeong Kim(Model)

서울 루이비통 메종 Seoul Louis Vuitton Maison

Location: Cheongdam, Seoul, Republic of Korea
Use: Commercial
Site Area: 938 m2
Building Area: 599 m2
Building Coverage ratio: 63.85%
Gross Area: 3,841 m2
Structure: Steel
Height: 24m
Photographer: Jaekyeong Kim(Model)

프레서울건축비엔날레 파빌리온 Seoul architecture biennale pavilion

Photographer: Junhwan Yoon, Unsangdong

전주 무형문화유산의 전당 Jeonju Intangible Cultural Heritage Hall

Location: 896-1 Dongsuhhak-dong Wansan-gu, Jeonju, Republic of Korea
Use: cultural facility
Site area: 63,020.00m²
Building area: 13,727m²
Gross floor area: 33,174.64m²
Building coverage: 21.78%
Floor space index: 37.51%
Building scope: B1 ~ 4F
Structure: Steel + Reinforced Concrete
Photographer: Junhwan Yoon, 토문엔지니어링 제공

SK 네트웍스 강남 사옥 SK Networks Gangnam Office

Location: Yeongdong-daero, Gangnam-gu, Seoul, Republic of Korea
Site area: 8,267.10 m²
Gross floor area: 47,307.83 m²
Photographer: Junhwan Yoon, Sergio Pirrone

다산동 성곽길 주차장 및 문화센터 Dasan-Dong Fortress Wall of Seoul Parking and Cultural Center

Location: Jung-gu, Seoul, Republic of Korea
Site area: 4,275.3m²
Gross floor area: 10,050.8m²
Building area: 2,896.9m²
Building to land ratio: 67.8%
Floor area ratio: 235.1%
Structure: Reinforced Concrete
Building scope: B3 ~ 3F
Height: 22m
Photographer: Jaekyeong Kim(Model)

갤러리 보고재 Gallery Vogoze

Location: 65-9 Samsung-dong, Gangnam-gu, Seoul, Republic of Korea
Site Area: 746 m²
Building Area: 371.89 m²
Total Floor Area: 3436.62m²
Client: Suwon Hong Gallery Vogoze
Structure: RC Structure
Architects: Jang Yoon Gyoo, Shin Chang Hoon
Designer: Kim Kyung Tae, Park Seong Yeon, Choi Yeong Eun, Kang Seung Hyun, Ko Eun JIn, Chang Chul Min, Sim Jehyun
Photographer: Jaekyeong Kim(Model), Sergio Pirrone

미동전자 사옥 Midong Electronics & Telecommunications Headquarter Office

Location: Yeoksam-dong, Gangnam-gu, Seoul, Republic of Korea
Site area: 949.9 m2 Program: Office, Shop
Building area: 473.33 m²
Building to land ratio: 49.80%
Gross floor area: 4,022.71 m²
Gross floor ratio: 285.70 %
Height: 24.50m
Architect: Unsangdong Architects
Design Team: Sangho Jeong, Samyeol Yoo, Seunghyun Kang, Soohoon Choi, Minkyun Kim
Design period: 2014.12~2015.04
Completion date: 2016.04
Photographer: Unsangdong(Model), Sergio Pirrone

청심유치원 White Quarter Circle

Location: 771-6, Yeoksam-dong, Gangnam-gu, Seoul, Republic of Korea
Use: Educational Facilities, Commercial
Site area: 577.50m²
Building area: 322.48m²
Gross floor area: 2,095.79m²
Building scope: B3 ~ 5F
Height: 26.61m
Building to land ratio: 55.84%
Floor area ratio: 199.83%
Structure: Reinforced concrete structure
Architects: Unsangdong Architects (Yoongyoo Jang, Changhoon Shin)
Design team: Gyeongtae Kim, Soohoon Choi, , Jeongseop Kim
Interior team: USD Design Group (Jaehyeon Shim, Byeongu Kim, Eunyeong Hwang)
Photographer: Jaekyeong Kim(Model), Yongkwan Kim, Jaeyoun Kim

우간다 힐링 마운틴(Healing Mountain) 청소년센터 UGANDA Healing Mountain

Photographer: Jaekyeong Kim(Model), Unsangdong

전곡리 선사박물관 Jeongok Prehistory Museum

Location: 528-1 Jeongok-ri, Yeonchun-gun, Gyeonggi-do, Republic of Korea
Architects: Jang Yoon Gyoo, Shin Chang Hoon
Designer: Kim Woo Young, Kim Yoon Soo, Kim Sung Min, Choi Hye Jin, Kwon Woo Seok, Jung Bok Joo
Photographer: Jaekyeong Kim(Model)

한내 지혜의 숲 Hannae Forest of Wisdom

Location: Seoul, Republic of Korea
Architects: Yoongyoo Jang, Changhoon Shin
Design team: Soohoon Choi
Client: Nowon-gu District Office
Completion: 2017.03
Gross Floor Area: 359.37m²
Photographer: Jaekyeong Kim(Model), Junhwan Yoon, Sergio Pirrone

이상봉 타워 Lie Sang Bong Tower

Location: 97-6, Cheongdam-dong, Gangnam-gu, Seoul, Republic of Korea
Use: Commercial, Office
Gross Floor Area: 5,007.34m²
Building scope: B5 ~ 14F
Height: 69.97m
Architects: Yoongyoo Jang, Changhoon Shin
Design team: Bongkyun Kim, Minseung Moon, Huijin Lee, Minkyun Kim, Taekmin Kim, Inhu Lim, Yeseul Jeon, Minji Sohn, Yuyeong Hwang
Structure: Reinforced concrete structure
Client: Lie Sangbong
Photographer: Jaekyeong Kim(Model), Jaeyoun Kim